DIGITAL REVOLUTION

GUIDE TO SUCCEED WITH TECHNOLOGY INTEGRATION INDUSTRY 4.0 INDUSTRIAL IoT

KAPIL KHURANA

ISBN
Paperback 979-8-89415-627-9
Hardcase 979-8-89446-692-7

Contents

Preface   *xi*

About Author   *xiii*

Acknowledgement   *xv*

Chapter 1: A Comprehensive Introduction: Industry 4.0   **1**

1.1 Industry 4.0   1

1.2 Industry 4.0 : Smart Factories   4

1.3 Advancement in Industries: Industry 4.0   6

1.4 IIoT: Industry 4.0 Prospect   8

1.5 How IIoT Will Drive the Fourth Industrial Revolution?   12

Chapter 2: Industrial Internet of Things (IIoT)   **20**

2.1 Introduction to IIoT   20

2.2 Evolution of IIoT Technology   22

2.3 Futuristic Nature of IIoT   24

2.4 Influence on Job Roles and Skill Requirements   25

2.5 Socio-Economic Implications of Widespread IIoT Adoption   26

2.6 Challenges and Considerations in IIoT Implementation   28

2.7 Emerging Technologies Shaping the Future of IIoT   29

2.8 Anticipated Advancements and Potential Disruptions in
     Industrial Automation   31

2.9 IIoT applications   33

Chapter 3: Basics of Communication Network: IIoT   **40**

3.1 Fundamentals of IIoT   40

3.2 Internet / WEB and Network Fundamentals OSI Model   42

3.3 Network access and physical layer IoT network technologies   44

3.4 IoT Networking Consideration and challenges   48

3.5 IoT Reference Model   51

3.6 Internet of Things Ecosystem   52

Chapter 4: Augmented /Virtual Reality: Industry 4.0 **72**

4.0 Introduction AR & VR 72

4.1 Process of Developing Augmented Reality Apps 76

4.2 Process of Developing Virtual Reality Apps 77

4.2 Best Practices for Augmented Reality App Development 79

4.4 New Trends in AR and VR Application Development 81

4.5 Enhancing Efficiency: AR/VR's Impact on Manufacturing 82

4.6 Beyond Boundaries: Exploring the Vast Potential of AR/VR 83

4.7 Creating Immersive Experiences: The Integration of AR/VR in Industry 4.0 Practices 85

4.8 Innovating Industry 86

Chapter 5: Blockchain Industry 4.0 **92**

5.1 Introduction Blockchain 92

5.2 Blockchain Unleashed: Transforming Industry 4.0 93

5.3 Decentralised Innovation: The Role of Blockchain in Industry 4.0 94

5.4 Building Trust: Blockchain's Impact on Industry 4.0 95

5.5 Securing the Future: Blockchain Technology in Industry 4.0 96

5.6 Beyond Transactions: Exploring Blockchain Applications in Industry 4.0 98

5.7 Data Integrity Revolution: How Blockchain is Reshaping Industry 4.0 99

5.8 Empowering Collaboration: Blockchain's Role in Industry 4.0 Ecosystems 100

5.9 The Expansion of Blockchain Applications in Industry 4.0 100

Chapter 6: Robotics in the Era of Industry 4.0 **106**

6.0 Basic Principles of Robotics 106

6.1 Hardware Components 107

6.2 Some of the critical advances in robotics include 108

6.3 Smart Manufacturing 109

6.4 Warehouse and Logistics   110

6.5 Healthcare and Life Sciences   111

6.6 Challenges and Considerations in the Integration of Robotics in Industry 4.0   113

6.7 Future Trends in Robotics for Industry 4.0   114

Chapter 7: Cloud Computing in Industry 4.0   **121**

7.1 Significance of Cloud Computing   121

7.2 Cloud Empowering Industry 4.0, Revolutionising Manufacturing and Beyond   124

7.3 Transitioning from Data Centres to Smart Factories: Cloud Computing in Industry 4.0   125

7.4 Embracing Cloud Innovation: Enabling the Future of Industry 4.0   125

7.5 Scalability and Flexibility: Cloud Computing's Role in Industry 4.0   126

7.6 Cloud-Driven Innovation: Transforming Industries in the Fourth Industrial Revolution   127

7.7 Virtualization and Connectivity: Cloud's Contribution to Industry 4.0   127

7.8 Cloud-Based Intelligence: Enhancing Efficiency in Industry 4.0 Systems   129

Chapter 8: Edge Computing in Industry 4.0   **137**

8.1 Significance of Edge Computing   137

8.2 Edge Computing: Pioneering the Next Era of Industry 4.0   138

8.3 Bringing Intelligence Closer: The Crucial Role of Edge Computing in Industry 4.0   140

8.4 Empowering Smart Factories: The Catalytic Role of Edge Computing in Industry 4.0 Success   143

8.5 Reshaping the Industrial Landscape: Edge Computing's Contribution to Industry 4.0   145

Chapter 9: Big Data Analytics and Software-Defined Networks: Industry 4.0 — **152**

9.1 Role of Data Analytics in Industry 4.0 — 152

9.2 Network Flexibility Software-defined networking (SDN) — 154

9.3 Information Pre-processing and Highlight Engineering — 156

9.4 Building and Preparing Models for IIoT Applications — 156

9.5 Artificial Intelligence Overview — 157

9.6 Big data — 159

9.7 Introduction to R and Julia Programming Languages — 163

9.8 Overview of Hadoop in the Context of IIoT — 165

Chapter 10: Integration of Different Technologies - Industry 4.0 — **171**

10.1 Advantages of Integration — 171

10.2 How do AR/VR, AI, big data, and advanced analytics interact? — 173

10.3 Challenges and Opportunities in Collaboration — 174

Chapter 11: Cyber Security - Industry 4.0 — **182**

11.0 Cyber Security — 182

11.1 Introduction to the Cyber Security Threat Landscape in Industry 4.0 — 184

11.2 Objectives of Business Management — 184

11.3 Vulnerabilities in Industry 4.0 Systems — 186

11.4 The role of employee training in cybersecurity — 189

11.5 Regulatory environment and compliance in industrial cyber security — 189

11.6 Changing threats in the business environment — 194

11.7 Inadequate encryption techniques — 195

11.8 Emerging Trends in IIoT Security — 197

11.9 Manufactured Insights and Machine Learning — 198

11.10 Edge Computing and Fog Computing: - IIoT Security — 198

11.11 Distributing Security Measures Closer to the Information Source — 199

11.12 Zero Trust Security Model - IIoT Security   199

11.13 Quantum-Safe Cryptography   200

11.14 How to progress IIoT security in industrial settings   201

11.15 Success Stories in Securing Factory Environments   201

11.16 Inadequate Network Security   209

11.17 The role of international standards in improving cybersecurity   213

Chapter 12: Digital Strategy Framework – Industry 4.0   **221**

12.0 Introduction   221

12.1 Meeting Customer Expectations   222

12.2 The Role of Governing Entities   222

12.3 Leveraging Third-Party Networks   222

12.4 Key Internal Forces Steering Industry 4.0 Integration   223

12.5 Level of Adoption   226

12.6 Does Every Company Need a Digital Strategy?   227

12.7 Production Automation   228

12.8 Process Reconstruction   229

12.9 Digital Platform Development   234

12.10 Challenges in Adopting Industry 4.0 Technologies   234

12.11 Role of Leadership and People to implement Digitization in Industries   235

12.12 Role of Investors and Boards in Digital Transformation   237

12.13 Chief Digital Officer   237

12.14 Profile structure and reporting   238

12.15 Talent acquisition   240

Chapter 13: Transformation from Industry 3.0 to Industry 4.0   **248**

13.1 Prepare for Industry 4.0   248

13.2 From Industry 3.0 to Industry 4.0   249

13.3 Innovation   251

13.4 Different Strategies to Implementation of Industry 4.0   253

13.5 Stages in the evolution of Industry 4.0 254

13.6 How to implement an Industry 4.0 strategy –
Implementing method 256

13.7 The Fourth Industrial Revolution: Shaping Global Trends 257

13.8 LEAN Production Systems 259

13.9 Review of IIoT Core Value Statements 263

13.10 The Role of IIoT in Shaping the Future of Industrial
Processes 264

Chapter 14: Industrial IoT Applications & Use Cases **270**

14.0 IIoT Industrial Applications 270

14.1 Key components of IIoT 270

14.2 Importance of IIoT in Changing Manufacturing Processes 271

14.3 Healthcare 272

14.4 Inventory Administration & Quality Control 276

14.5 Building Automation 281

14.6 Security and Surveillance 282

14.7 Application Domains: Oil, Chemical and
Pharmaceutical Industry 283

Glossary 299

Reference 303

Preface

Welcome to "Digital Revolution," a comprehensive guide that unravels the intricacies of the transition to Industry 4.0. In this book, we take you on a journey through the industry's evolution, tracing its roots from the pages of history to the forefront of the digital age. From the advent of steam power to the dawn of Industrial IoT, we explore the milestones that have shaped the industrial landscape and paved the way for the transformative changes we witness today, ensuring you have a comprehensive understanding of Industry 4.0.

"Digital Revolution" is not just a chronicle of technological advancements; it is a practical roadmap for embracing the industry's future. Through each phase of the industrial revolution, we delve into the key concepts, methodologies, and technologies that have propelled progress forward. From the early days of mechanisation to the advent of cyber-physical systems, we examine the convergence of physical and digital realms and its profound implications for businesses and society, equipping you with the knowledge and tools to navigate these changes.

Central to our exploration is the theme of automation and integration—a cornerstone of Industry 4.0. We dissect the various facets of automation, from robotics and artificial intelligence to data analytics and machine learning, offering insights into their applications and potential impact on production processes. However, it's important to note that implementing these technologies can come with challenges. For instance, companies may face resistance from employees who fear job loss due to automation. We discuss strategies for addressing these challenges and opportunities of rapid technological advancement, such as retraining employees for new roles or focusing on tasks that require human creativity and problem-solving skills.

Beyond the technical realm, "Digital Revolution" delves into the broader implications of digital transformation, exploring its ramifications for careers, industries, and global economies. We examine the role of protocols and standards in facilitating interoperability and seamless integration across disparate systems. Furthermore, we discuss the strategic imperatives for organisations seeking to thrive in an increasingly interconnected and competitive landscape. However, it's crucial to consider the ethical implications of these technologies. For instance, the use of artificial intelligence in decision-making processes can raise questions about fairness and bias. We encourage readers to think critically about these issues and consider how they can be addressed.

As you begin on this journey through the pages of "Digital Revolution," I invite you to embrace curiosity and open-mindedness. Whether you are a seasoned industry professional or an aspiring technologist, this book offers valuable insights and practical guidance for navigating the complexities of Industry 4.0. For instance, we could provide tips on how to stay updated with the latest technological advancements or how to foster a culture of innovation within an organisation. Together, let us start on a transformative odyssey that promises to reshape the future of industry and society as we know it.

About Author

Kapil Khurana is a proficient in IIoT (Industrial Internet of Things) and Industrial Automation. With over 18 years of experience in industry, Kapil Khurana has been at the forefront of technological innovations shaping the future of manufacturing and industrial processes.

Kapil Khurana has translated his expertise into practical solutions. His research and implementations in IIoT systems and industrial automation have significantly advanced smart factories, predictive maintenance strategies, and real-time data analytics in industrial settings.

Kapil Khurana's contributions to the field of IIoT and Industrial Automation have not gone unnoticed. His ability to simplify complex concepts in his numerous peer-reviewed papers and technical articles has earned him recognition as a thought leader in the industry.

Beyond his academic pursuits, Kapil Khurana has collaborated with leading industrial players, implementing advance IIoT solutions to optimize production processes, enhance operational efficiency, and ensure seamless integration of digital technologies with traditional manufacturing systems.

With heartfelt gratitude,

Kapil Khurana

Acknowledgement

To my beloved parents, Mr. J.D. Khurana & Mrs Bhagwanti Khurana, your love and unwavering faith have been the foundation of everything I have achieved.

Rashmi, you are the most amazing companion that I have been blessed with. Thanks would not even begin to justify for everything you do every single day to make our life awesome.

Amit & Sneha, your warmth, affection and unwavering belief in me has given me the courage to pursue my dreams and overcome every obstacle.

Kairavi & Jaisvi your unconditional love, priceless laughter and boundless energy fills my heart with joy and gratitude every single day.

Anuu Priya, thank you for being an incredible sister and an indispensable part of this project. Your encouragement and insightful advice have been invaluable throughout this journey.

I would like to acknowledge and extend a very special thanks to Mr. Sandeep Arora, Chancellor – Kalinga University, who not only provided the resources but also shared his knowledge, connecting me with the experts in the field that I needed to bring this vision to life. I will always be eternally grateful for your guidance, support, boundless generosity, and enduring friendship.

Also, would like to thanks Mr. Pranshu Arora, Mr. Hani Dixit, Ms Archana Bhakta, Ms Shalu Gupta & Mehak Sharma for their help.

Last but not the least, you, the reader. I hope that through my book I was able to add some value to your quest for learning and growing.

With heartfelt gratitude,

Kapil Khurana

Chapter 1

A Comprehensive Introduction: Industry 4.0

Learning Objectives:

- Understand the concept of Industry 4.0.

- Identify the key technological advancements driving Industry 4.0.

- Examine the role of the Industrial Internet of Things (IIoT) in Industry 4.0.

- Understand the architecture components and security considerations associated with IIoT implementation.

- Industrial IoT: Business Model and Reference Architecture.

1.1 Industry 4.0

Industry 4.0, also known as the Fourth Industrial Revolution, is all about integrating new digital technologies like the Internet of Things (IoT) into traditional manufacturing processes. It started in Germany but has now spread worldwide to places like Europe, India, and China. Basically, it means connecting up all your factory equipment and assets using cyber-physical systems. This allows them not only to share info but also control each other to become smarter.

Fourth Industrial Revolution changing everything through technologies like AI, robotics, the internet of things (IoT), and cloud computing.

This is transforming every part of our lives from how we do our jobs to how we connect with each other. While it creates opportunities for innovation and economic growth, it also brings challenges like jobs being replaced and needing new skills.

Even though we're still in early stage, but it's already having a huge impact worldwide. To prepare for the future, it's important we understand what this fourth revolution is and how it's altering our lives.

To properly understand the Industry 4.0, we must first examine the transformations and historical context of the different industrial revolutions and their impact on society.

Industrial transformation is not a new phenomenon, industries have evolved themselves repeatedly over the past centuries. Individual sectors transformed due to new technologies, discoveries, inventions, or processes becoming available.

On a macro level, businesses and industries have progressed through at least four stages of change over the last centuries. These stages are commonly referred to as the four industrial revolutions. It is essential to understand the prior industrial revolutions as they provide context for the current transformation known as the Fourth Industrial Revolution.

Below are the details about the four Industrial Revolutions from Industry 1.0 to Industry 4.0:

Industry 1.0 - Mechanization and Steam Power (Late 18th to Early 19th Century): The first industrial revolution marked a pivotal shift from agrarian economies to industrial societies. It was characterised by the mechanisation of production processes, driven primarily by innovations such as the steam engine and mechanical loom. Steam power enabled factories to mechanise tasks that were previously done by hand, leading to unprecedented increases in productivity and the establishment of large-scale manufacturing facilities. This era laid the foundation for mass production and the mechanisation of labour, fundamentally altering the economic and social landscape.

Industry 2.0 - Mass Production and Electrification (Late 19th to Early 20th Century): Building upon the innovations of the first industrial revolution, the second industrial revolution brought about the widespread adoption of electricity and assembly-line manufacturing. Pioneered by visionaries such as Henry Ford and Thomas Edison, mass production techniques revolutionised industries ranging from automotive to consumer goods. Electrification enabled factories to operate more efficiently, powering machines and equipment with greater reliability and precision. This era saw the rise of standardised products, economies of scale, and the emergence of modern corporate structures. It laid the groundwork for the concept of interchangeable parts, specialisation of labour, and the division of tasks—a precursor to the efficiency-driven ethos of Industry 4.0.

Industry 3.0 - Automation and Computerisation (Mid-20th Century to Late 20th Century): The third industrial revolution ushered in the era of automation and computerization, driven by advancements in electronics, telecommunications, and digital technology. Key innovations such as transistors, microprocessors, and the internet transformed industries, enabling greater levels of automation and connectivity. Computer-controlled systems revolutionised production processes, enabling programmable logic controllers

(PLCs) and robotics to automate tasks with unprecedented speed and precision. This era also saw the emergence of computer-aided design (CAD), computer-aided manufacturing (CAM), and enterprise resource planning (ERP) systems, laying the groundwork for the integration of digital technologies into industrial workflows.

Industry 4.0 is built upon a suite of cutting-edge technologies that collectively empower organisations to create interconnected, intelligent, and autonomous systems. These foundational technologies form the backbone of Industry 4.0, enabling the digital transformation of industrial processes and unlocking new levels of efficiency, agility, and innovation.

Industry 4.0 focuses on optimising smart, flexible supply chains, factories and distribution through machine-to-machine communications and data sharing with human operators. This aims to enable faster, smarter decisions while minimising costs. Ultimately changing company and regional competitiveness, it increases manufacturing productivity, shifts economics and modifies workforces for faster industrial growth.

Industrial Revolution	Period	Key Features
Industry 1.0	Late 18th - Early 19th Century	Mechanisation with water and steam power; Introduction of the mechanised loom and steam engine.
Industry 2.0	Late 19th - Early 20th Century	Mass production through assembly lines; Electricity, internal combustion engine, and telegraph communication.
Industry 3.0	Mid 20th Century	Automation and computerization; Adoption of electronics and IT systems, including computers and the internet.
Industry 4.0	Present	Cyber-physical systems; Integration of IoT, AI, machine learning, big data, cloud computing, and robotics.

Exhibit 1.1 industry 4.0 revolution

1.2 Industry 4.0 : Smart Factories

Lots of innovations are happening with smart factories these days. Factories are getting way more high-tech with the rise of things like smart machines and intelligent products.

The core of Industry 4.0 is the smart factory. In a smart factory, not only the production processes and machinery be intelligent, but the products being made will have smarts built right in too. That means the products will actually carry information about themselves like what they are, when they were made, and where they are currently. Pretty wild, right?!

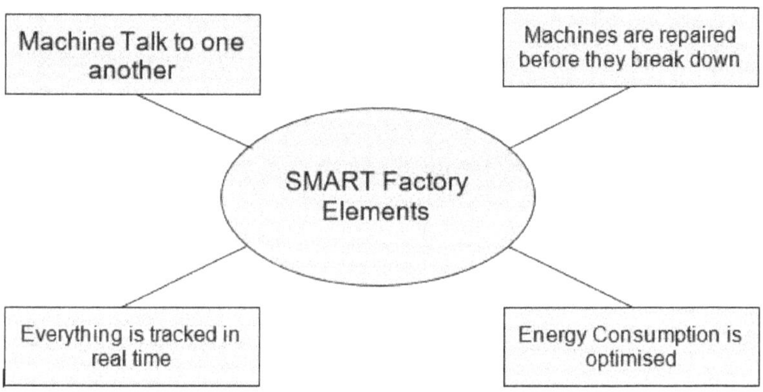

Exhibit 1.2 Smart Factory

They'll even know what steps are left to complete them. The products will hold details about vendors, operators, and quality that help with traceability and access. That embedded info lets the products suggest optimised routes and methods for finishing the production process. A smart product could literally tell the transport system which path to take to wrap things up.

Another big part of Industry 4.0 is integrating the "horizontal" business operations like manufacturing with the "vertical" ones like finance, HR, sales, logistics and more. The goal is end-to-end process management from the supply chain all the way through services and the product life cycle. Basically, merging information tech with operational tech so that it all works seamlessly as one unified system. Pretty ambitious but it has the potential to seriously upgrade how factories operate.

Industry 4.0 is suitable not just for large companies but also for small and medium-sized enterprises. Flexibility and control over the manufacturing process enabled by smart products like RFID systems allow manufacturers to make better decisions with dynamic process control and customization capabilities. Smart factories provide the ability to easily change product designs and variants quickly without requiring major capital investments or machinery changes. Dynamic business and engineering processes mean small and medium enterprises can also become competitive by enabling new ways of creating value and innovative business models.

By utilising various advanced technologies, Industry 4.0 provides improvements in production processes, communication, and information technology that help increase automation and digitization across entire manufacturing processes.

The Fourth Industrial Revolution is the ongoing automation and digitalization of traditional manufacturing practices using modern innovative technologies. Large-scale machine-to-machine communication and the Internet of Things are integrated for increased automation, improved communication, self-monitoring, and the production of intelligent products. This transformative revolution impacts all aspects of life from how we work to how we live.

Industry 4.0 represents the convergence of disruptive digital technologies that will fundamentally change manufacturing beyond what is imaginable today. Driven by extraordinary increases in data velocity, volume, and variety as well as system integrations, connectivity, advanced analytics, business intelligence capabilities, machine learning, and improvements transferring digital instructions to the physical world. Companies may pursue efforts to digitise factories. Among other impacts, introducing Industry 4.0 can lead to sensor-equipped, internet-connected products providing better products and services to customers.

Understanding the implications of the Fourth Revolution for businesses and individuals is essential. It drives increased efficiency and productivity across all business sizes. By integrating digital and physical technologies, companies can automate tasks, reduce waste, and improve overall performance. It creates new opportunities for innovation and growth as businesses harnessing this system can develop new products and services, and expand into new markets.

Industry 4.0 is changing the way we work. As more tasks are automated, workers will need to develop new skills and adapt to new working models.

Aspect	Description
Smart Factory	Factories incorporating smart machines and intelligent products, where production processes and machinery are intelligent. Products carry information about themselves.
Smart Products	Products with embedded information regarding their identity, production details, and current status.
Traceability	Products holding details about vendors, operators, and quality to enhance traceability and access.
Optimization	Embedded information in products allows for suggesting optimised routes and methods for production processes.
Integration	Integration of horizontal and vertical business operations for end-to-end process management.

Exhibit 1.3 Industry 4.0 Key Elements

1.3 Advancement in Industries: Industry 4.0

Industry 4.0 represents a paradigm shift fuelled by the integration of digital technologies into industrial processes. Industry 4.0 is characterised by the convergence of physical and digital systems, creating intelligent, interconnected, and data-driven manufacturing environments.

Industry 4.0 leverages big data analytics to process vast amounts of data generated by industrial processes. This enables organisations to gain actionable insights, optimise production, and make data-driven decisions.

The incorporation of artificial intelligence (AI) and machine learning (ML) further enhances the capabilities of Industry 4.0. These technologies enable systems to learn and adapt, optimising operations and predicting maintenance needs.

In Industry 4.0, Smart factories do not work on a standalone basis. There is a communication network through which data is transmitted between Smart Factories to smart production systems to smart machines to smart products. This is implemented through Cyber Physical Production Systems (CPPS), which lets factories react quickly and appropriately to variables, such as demand levels, stock levels, machine defects and unwanted breakdowns.

By integrating different elements like Logistics and marketing services can make the overall supply chain and target market actions quicker and more dynamic.

Integration across the globe or outside the production premises, that means integration between the business associates and the customers. However, it could also mean the integration of business models across countries and even across continents, making for a global network.

The complete lifecycle of the product is traced from production to retirement. Major manufacturing Industries focus on the manufacturing process, to make the product, sell the product, and ship it. They don't have any feedback mechanism through engineering to get the analysis data about the performance of the product. If quality is not good then it will impact future sales trends. However, when dealing with industrial products, quality and customer satisfaction matters a lot. Industry 4.0 covers both the production process and the entire lifecycle of the product.

A closer look at industry 4.0 reveals a powerful confluence of management trends and emerging technologies that promise to change the way factories work. Everyone agrees, the impact of this change will be unprecedented, affecting every industry in every country and transforming entire product lines, production systems, business models, and service delivery formats. However, according to the 2018 World Economic Forum report The Next Economic Growth Engine: Scaling Fourth Industrial Revolution Technologies in Production, "More than 70 percent of industrial companies are still either at the start of the journey or unable to go beyond the pilot stage." Clearly, we are still at the cusp of this revolution.

So, what is stopping Industry 4.0?

A closer look at the implementation approach will help answer this question. At the core of this Industry 4.0 are cyber – physical systems, the coming together of diverse technologies, and the intertwining of virtual and the physical worlds. Although the effects of this change will be seen as revolutionary when considered in retrospect, for most organisations the transformation will be gradual – an evolution rather than a revolution.

Industry 4.0 is not a greenfield project for most companies, unless they happen to be a digital upstart like Tesla. Technology executives need to maintain continuity with legacy systems that have evolved over the decades and are still functional.

Many enterprises have started pilots to understand how Industry 4.0 can be planned and implemented. Unfortunately, there are not many instances of these pilots scaling up to large – scale implementations, for a variety of reasons. The people element has been widely reported as a leading cause – A systematic approach with clear and specific milestones which are measurable and visible can successfully implement and work.

1.4 IIoT: Industry 4.0 Prospect

Industrial Internet of Things (IIoT) is closely related to the Industry 4.0 revolution, which represents the fourth industrial revolution characterised by the integration of digital technologies into production and industrial processes. Industry 4.0 emphasises the use of smart, connected systems to create a more responsive and efficient industrial ecosystem.

The Industrial Internet of Things (IIoT) represents the intersection of traditional industrial processes with advanced technologies and creates a network of interconnected devices and systems. Unlike customer-focused IoT, IIoT is tailored for industrial applications to improve efficiency, productivity and decision-making in manufacturing and other industries.

IIoT relies on the connectivity of devices, sensors and machines within an industrial ecosystem. These components communicate and share data to enable intelligent decision making.

IIoT involves the integration of data from various sources and provides a holistic view of industrial processes. This data-driven approach facilitates real-time monitoring, analysis and optimization.

IIoT uses automation to streamline industrial workflows. By connecting devices and systems, it enables autonomous decision-making and increases operational efficiency.

Key Features	Description
IoT Integration	Interconnected devices and sensors collect real-time data for monitoring and optimization.
AI and Machine Learning	Analyses data to predict maintenance needs, optimise production schedules, and improve efficiency.
Robotics and Automation	Automated systems handle repetitive tasks, increasing production speed and reducing errors.
Cloud Computing	Centralised data storage and processing enable remote access, collaboration, and scalability.
Digital Twins	Virtual replicas of physical assets allow for simulation, testing, and optimization of processes.

Exhibit 1.4 Technology Drivers

The importance of IIoT lies in its transformative impact on traditional industrial processes, offering a number of benefits like:

Operational efficiency: IIoT enables real-time monitoring of equipment and processes, enables immediate response to deviations, and minimises downtime. Predictive maintenance based on data insights ensures that machinery is operating at peak efficiency.

Data-driven decision-making: With the large amount of data generated by connected devices, IIoT enables decision-makers to gain actionable insights. This data-driven approach facilitates informed decision-making, leading to improved outcomes and resource utilisation.

Cost reduction: IIoT helps optimise resource utilisation, reduce waste and increase overall operational efficiency. Predictive maintenance and condition monitoring contribute to cost savings by preventing costly equipment failures.

Innovation and Agility: By supporting connectivity and collaboration, the IIoT supports innovation in industrial processes. It enables the adoption of new technologies, making industries more adaptable and agile in response to market changes.

1.4.1 Industrial IoT: Business Model and Reference Architecture

As of now we discussed about the basic concepts of Industry 4.0 and how IIoT is adding value to achieve these concepts in SMART Factories. Now let's discuss about the different business models which is generally used while outsourcing the IIoT services. At the end of the chapter you can decide which model suits you the most:

Subscription-based Models: In a subscription-based model, industrial users pay a recurring fee to access and use IIoT services. This model provides users with continuous access to IIoT features for a specified period, similar to subscribing to a service. Predictable revenue is generated through recurring subscriptions which offer a steady and predictable stream of revenue for IIoT service providers. Subscribers also benefit from continuous updates and improvements to the IIoT platform throughout their subscription period.

Pay-per-Use Models: Pay-per-use models involve users paying for IIoT services based on their actual usage. This could include factors such as the volume of data processed, the number of devices connected, or the frequency of specific features utilised. This model is cost-efficient and scalable as users only pay for the resources they consume. It is also flexible as costs align with actual usage, making it ideal for industries with varying workloads.

Outcome-based Models: In an outcome-based model, the pricing is tied to the outcomes or results achieved by the industrial user. This could include improved efficiency, reduced downtime, or other predefined key performance indicators (KPIs). This model aligns the interests of IIoT service providers with the success and performance improvements of the industrial users. It also enables risk sharing as providers share the risks and rewards based on the actual outcomes achieved.

Platform as a Service (PaaS): PaaS involves providing a comprehensive IIoT platform that users can leverage to build, deploy, and manage their own IIoT applications. Users pay for the platform and services they use, typically on a subscription basis. This provides development flexibility as users have the ability to develop and deploy their custom IIoT applications on the platform. It also yields cost savings by reducing the need for users to build their IIoT infrastructure from scratch.

Data Monetization Strategies: Data monetization involves extracting value from the data generated by IIoT devices. This can be achieved by selling data to third parties, creating data-driven products, or offering insights derived from the data as a service. This enables IIoT providers to generate revenue not only from their primary users but also by monetizing the valuable data they collect. It also allows IIoT providers to tap into new markets beyond their core industrial users.

Model	Description	Advantages
Subscription-based	Users pay a recurring fee for continuous access to IIoT services.	Predictable Revenue - Continuous Updates
Pay-per-Use	Users pay based on actual usage, such as data volume or device connections.	Cost Efficiency - Flexibility
Outcome-based	Pricing tied to achieved outcomes like efficiency improvements or reduced downtime.	Performance Alignment - Risk Sharing
Platform as a Service	Provides a comprehensive IIoT platform for users to build and manage applications.	Development Flexibility - Cost Savings

Model	Description	Advantages
Data Monetization	Extracting value from IIoT data through selling data, creating products, or offering insights.	Diversified Revenue Streams - Market Expansion

Exhibit 1.5 IIoT Business Models

1.4.2 Basic Architecture Components for IIoT

The reference architecture includes essential components like edge computing, cloud integration, sensor networks, data storage, and security measures to create a comprehensive and secure IIoT ecosystem.

Edge Computing in IIoT Architecture

Edge computing is a pivotal component of IIoT architecture that involves processing data closer to where it is generated rather than solely relying on centralised cloud servers.

Edge computing minimises data transfer time, enabling real-time decision-making essential for industrial processes. By processing data locally, edge computing reduces the need for large data transfers over the network, optimising bandwidth usage. Edge devices can make autonomous decisions based on local data, enhancing efficiency and responsiveness.

Cloud Integration and Services

Cloud integration is core in IIoT architecture, providing scalable storage, computing power, and a platform for advanced analytics.

Cloud services enable analysis of vast amounts of IIoT-generated data, extracting valuable insights for decision-making. Cloud platforms offer scalability, allowing IIoT systems to handle varying workloads and accommodate growth of connected devices. Cloud-based services facilitate collaboration by providing a centralised platform for multiple stakeholders to access and analyse data.

Sensor Networks and Device Connectivity

Sensor networks and device connectivity form the foundation of the IIoT ecosystem, collecting and transmitting data from physical assets.

Sensors collect real-time data from machines, equipment, and infrastructure, providing insights into operational conditions. Standardised protocols enable seamless communication between diverse devices, ensuring interoperability in the IIoT landscape. Device connectivity allows for remote monitoring of assets, enhancing visibility and control over industrial processes.

Data Storage and Management

Storage and management of vast amounts of data generated by IIoT devices are essential for historical analysis, reporting, and compliance.

Cloud-based storage solutions provide scalability required to handle massive volumes of data generated by IIoT devices. Define policies for storing and archiving data based on regulatory requirements and needs of industrial processes. Develop and implement data lifecycle management techniques.

Security and Privacy Considerations

Given the critical nature of industrial processes, robust security and privacy measures are essential to safeguard IIoT systems.

Implement end-to-end encryption to protect data during transmission and storage, preventing unauthorised access. Establish strict access controls to ensure only authorised personnel can interact with and manage IIoT systems. Implement continuous monitoring systems for anomaly detection, identifying and responding to security threats in real-time.

1.5 How IIoT Will Drive the Fourth Industrial Revolution?

The Fourth Industrial Revolution is a significant period in human history and is characterised by the convergence of the digital and physical world. At the heart of this transformative era is the Internet of Things (IIoT), a network of interconnected devices that communicate and exchange data, enabling unprecedented traffic across industries. As we read this chapter, we will understand the far-reaching impact of the Internet of Things and explore its role in the evolution of business, commerce and the fabric of our daily lives.

IoT or Industrial Internet of Things is a concept. The term refers to the Internet of Things (IIoT), a network of physical, automotive, home, and other devices with electronic components, software, sensors, actuators, and network connections that allow these devices to collect and exchange data. The Internet of Things could revolutionise many industries, including manufacturing, healthcare, transportation and agriculture. It helps increase efficiency, productivity and safety and develop new products and services. The Industrial Internet of Things (IIoT) is a key enabler of the fourth revolution. IIoT refers to a network of physical devices and other devices with sensors, software, and network connections that collect and share data. IIoT is used in many sectors to increase efficiency, productivity and security. For example, in manufacturing,

IIoT is used to monitor and improve production lines; to track information and improve diagnosis and treatment in healthcare; and to improve traffic management and reduce accidents in transportation.

How IIoT supporting Industry 4.0

Predictive maintenance: Industrial IoT can be used to collect data from machines and equipment to identify potential problems before they occur. This prevents cost reduction and improves asset utilisation.

Remote monitoring: IIoT can monitor and control devices and systems. This can be used to increase efficiency and productivity and ensure safety.

Instant Data Analysis: IIoT can collect and analyse data instantly. This information can be used to make better operational decisions, such as optimising production plans or adjusting equipment connections.

Self-healing: Industrial IoT can be used to create self-healing systems that can learn and adapt to their environment. It increases efficiency, productivity and reduces costs.

IIoT is still in its infancy but has the potential to revolutionise many industries. As IIoT technology evolves, it may play a significant role in the Fourth Generation.

IIoT is used to create new products and services. For example, smart homes that support the Industrial Internet of Things can adjust the temperature, and smart factories that support the Industrial Internet of Things can improve production processes.

The Fourth Industrial Revolution built on the foundations of its predecessors, with all the changes leading to advancements in energy technology. Leading mechanisation and steam power. The second pioneers mass production and electricity, while the third shows computers and automation. We are now witnessing the merging of digital, physical and virtual worlds in the Fourth Industrial Revolution, and the lines between the virtual and the visual are blurring.

Foundation At the heart of the Fourth Industrial Revolution is the Internet of Things, a concept that goes beyond connectivity. It covers a broad network of devices equipped with sensors, actuators and communication capabilities, ranging from everyday appliances to complex industrial machines. These connected devices create and share data, paving the way for a new era of intelligent decision-making and automation.

The impact of IIoT is impacting across all industries, changing their operations. Industrial IoT in smart homes facilitates the connection of devices, giving homeowners unprecedented control over their environment. From thermostats that adjust to user preferences to refrigerators that produce products, the integration of Industrial IoT into home environments is becoming easy and efficient. The fourth industrial revolution in manufacturing can be summarised with the rise of smart factories. IIoT-enabled sensors on machines collect real-time data to streamline production processes, predict maintenance needs and reduce downtime. This increases efficiency and transforms traditional manufacturing into a flexible, data-driven ecosystem.

Economic Transformation

The economic impact of the Fourth Industrial Revolution is far-reaching. Industrial IoT's potential to increase productivity and efficiency can translate into business growth. As the economy becomes more connected and data-driven, new business models are emerging that drive innovation and competition. Digitalization of the economy has become a support for job creation and skills development when there is a problem with traditional work patterns.

Industrial IoT in Healthcare and Agriculture

The impact of Industrial IoT is especially evident in healthcare, where it has revolutionised patient care and management. Wearable devices with health monitoring sensors can track vital signs. Available information improves prevention, helps create personalised treatment plans, and improves health outcomes.

In agriculture, precision farming driven by the Internet of Things has become a game changer. Sensors in the field collect data on soil conditions, weather conditions and crop health, allowing farmers to make data-driven decisions. This not only increases efficiency but also reduces resource consumption, thus solving increasing problems related to food security and sustainable agriculture.

Life and work in the age of the Industrial Internet of Things

The fourth industrial revolution has changed the nature of our lives and work. Smart cities are equipped with IoT infrastructure to effectively manage resources, improve public services and improve the overall quality of urban life. Employees are experiencing a transition to remote and flexible working, driven by Industrial IoT connectivity that transcends geographical boundaries.

As we march towards the dawn of the Fourth Industrial Revolution, it is clear that the transformative power of the Internet of Things goes beyond technological innovation. It redefines our social structures, economic systems, and the relationship between people and technology. In the next section, we will look more deeply at specific practices, issues, and ethical considerations during this period, illustrating the complexity of the world that emerges from the interaction between the digital and the physical.

Key Notes:

Topics	Keynotes
Industry 4.0	Integration of digital technologies into traditional manufacturing processes.
	Real-time monitoring and predictive maintenance capabilities enabled by IoT.
	Transition from isolated automation to connected, data-driven ecosystems.
Significance	Historical context of industrial revolutions provides insights into Industry 4.0.
	Transformation of businesses through technological advancements over centuries.
SMART Factories	Introduction of smart machines and intelligent products in manufacturing.
	Products with embedded information contribute to traceability and optimization.
Accessibility	Industry 4.0 benefits extend to small and medium-sized enterprises.
	Flexibility and control over manufacturing processes empower SMEs.
Integration	Integration of digital technologies like AI, ML, and IoT in industrial processes.
	Smart factories communicate through Cyber Physical Production Systems (CPPS).
Adoption Challenges	Significant portion of industrial companies are at the early stages of Industry 4.0 adoption.
	Identifying and overcoming barriers is essential for successful implementation.
IIoT Importance	IIoT enhances operational efficiency, data-driven decision-making, and innovation in industrial processes.
	Cost reduction and agility are among the key benefits of IIoT implementation.

Topics	Keynotes
IIoT Technologies	Sensors, actuators, wireless sensor networks, cloud computing are key components of IIoT.
	IIoT business models include subscription-based, pay-per-use, and outcome-based pricing models.
IIoT Architecture	Edge computing, cloud integration, sensor networks, and security measures are integral components of IIoT.
Security & Privacy	End-to-end encryption, access controls, and continuous monitoring are essential for IIoT security.
Impact of IIoT	IIoT plays a crucial role in driving the Fourth Industrial Revolution.
	Connectivity and data exchange facilitated by IIoT revolutionize various industries and daily life.

Trivia Challenge Industry 4.0:

Welcome to the Industry 4.0 Trivia Challenge! Get ready to test your knowledge of the wacky world of Industry 4.0 with these hilarious multiple-choice questions. Each question is paired with some side-splitting facts and figures about Industry 4.0. Let us dive in and have a giggle!

1) What is the significance of Industry 4.0?

A) It emphasizes traditional manufacturing practices.

B) It optimizes manufacturing processes through innovative technologies.

C) It focuses on maintaining manual labour.

D) It excludes small and medium-sized enterprises because they're just too small to handle the big tech stuff!

2) How does Tesla's Gigafactory utilize Industry 4.0 technologies?

A) By relying solely on manual labour……!!

B) By streamlining production processes and making cars at the speed of light...almost.

C) By avoiding technological advancements and sticking to the good old days of manufacturing.

D) By neglecting the use of automation.

3) What is the aim of Industry 4.0 technologies beyond large enterprises?

A) To limit accessibility to advanced technologies.

B) To democratize Industry 4.0 technologies and make them as familiar.

C) To exclude small and medium-sized enterprises because they are not cool enough.

D) To prioritize traditional manufacturing practices.

4) In which industries do advancements facilitated by Industry 4.0 technologies occur?

A) Healthcare and agriculture only.

B) Healthcare, agriculture, and automotive because innovation knows no bounds!

C) Automotive and aerospace only.

D) Information technology and finance because even bankers need a little tech love.

5) What challenges are inherent in implementing Industry 4.0 technologies?

A) None, as implementation is seamless.

B) Limited adoption due to technological complexity, but who does not love a good challenge?

C) Easy integration with existing systems.

D) Low investment requirements.

6) What aspect of Industry 4.0 do readers explore within the context of IIoT?

A) Business models and reference architectures.

B) Traditional manufacturing practices.

C) Exclusively large enterprise applications.

D) Manual labour techniques.

7) What do SMART factories aim to achieve?

A) Increased reliance on manual labour.

B) Optimized manufacturing processes through automation, data analytics, and interconnected machinery.

C) Reduction in technological advancements.

D) Exclusion of automation because efficiency is overrated.

8) How does Industry 4.0 impact traditional manufacturing practices?

A) By emphasizing manual labour.

B) By neglecting technological advancements.

C) By optimizing processes through innovative technologies.

D) By limiting accessibility to advanced technologies.

9) What is the broader implication of IIoT in industrial settings?

A) Exclusively for large enterprises.

B) Democratization of advanced technologies.

C) Maintenance of traditional manufacturing practices.

D) Reduction in efficiency!!

10) What do readers gain insights into regarding IIoT within Industry 4.0?

A) Business models and reference architectures!!

B) Limitations of IIoT implementation.

C) Exclusion of IIoT from industrial settings.

D) Dependence on manual labour techniques.

Summary:

Industry 4.0, the Fourth Industrial Revolution, revolutionizes traditional manufacturing by integrating digital technologies like IoT. Real-time monitoring and predictive maintenance enhance efficiency and reduce downtime. Smart factories and products contribute to traceability and optimization. Industry 4.0's significance lies in its historical context and its potential to empower small and medium-sized enterprises. However, adoption challenges persist, with many companies still at early stages. The Industrial Internet of Things (IIoT) plays a pivotal role in Industry 4.0, enhancing operational efficiency, decision-making, and innovation. IIoT's architecture involves edge computing, cloud integration, and robust security measures. Its impact extends beyond industries, driving the Fourth Industrial Revolution by enabling connectivity and data exchange.

Abbreviations:

Abbreviations	Terms
IIoT	Industrial Internet of Things
IoT	Internet of Things
CPS	Cyber Physical Systems
ML	Machine Learning
AI	Artificial Intelligence
SMEs	Small and Medium-sized Enterprises
RFID	Radio Frequency Identification
CPPS	Cyber Physical Production Systems
PaaS	Platform as a Service
4IR	Fourth Industrial Revolution
EDGE	Edge Computing
DDoS	Distributed Denial of Service
SDK	Software Development Kit
AM	Additive Manufacturing
MES	Manufacturing Execution System

Chapter 2

Industrial Internet of Things (IIoT)

Learning Objectives:

- Understand the concept of Industrial Internet of Things (IIoT)and its significance in modern industrial operations.

- Trace the evolution of IIoT technology from its inception to its current state, including key advancements and integrations.

- Explore the futuristic nature of IoT, including predictions about Its impaction industrial systems, digital twins, cyber-physical systems, and its influence on human roles.

- Analyse the influence of IIoT adoption on job roles and skill requirements, identifying the emerging roles and competencies needed in industrial settings.

- Examine the improvements in workplace safety and working conditions facilitated by IIoT technologies.

2.1 Introduction to IIoT

In recent years, the Industrial Internet of Things (IIoT) has emerged as a revolutionary force reshaping the landscape of industrial operations. At its core, IIoT refers to the interconnected network of smart devices, sensors, machines, and systems within industrial environments, all of which communicate and exchange data in real-time. This connectivity enables unprecedented levels of automation, efficiency, and insight into industrial processes.

IIoT represents a paradigm shift in industrial operations, moving away from traditional, siloes systems towards interconnected ecosystems that seamlessly integrate physical machinery with digital technologies. Unlike its predecessor, the Internet of Things (IoT), which primarily focuses on consumer applications, IIoT is tailored specifically for industrial settings, where its impact is profound and far-reaching.

The significance of the Industrial Internet of Things (IIoT) in the industrial context is indeed monumental, fundamentally reshaping how organisations operate and compete in today's digital age. At its core, IIoT harnesses the power of connectivity and data analytics to revolutionise industrial processes, offering unprecedented levels of optimization and efficiency. One of the key advantages of IIoT lies in its ability to enable real-time monitoring and control of equipment. By equipping machinery and assets with sensors and connectivity, organisations can gather real-time data on performance, condition, and usage, allowing for proactive maintenance and intervention to prevent costly downtime and equipment failures. This predictive maintenance capability not only reduces operational disruptions but also extends the lifespan of assets, optimising resource utilisation and minimising maintenance costs.

Moreover, IIoT facilitates the optimization of production processes by providing granular insights into performance metrics, such as throughput, yield, and energy consumption. By analysing data collected from sensors and devices throughout the production line, organisations can identify inefficiencies, bottlenecks, and opportunities for improvement, enabling them to fine-tune operations for increased efficiency and quality. This data-driven approach to process optimization enables organisations to achieve higher levels of productivity while maintaining stringent quality standards, ultimately driving competitive advantage in the marketplace. By leveraging advanced analytics and machine learning algorithms, organisations can derive actionable insights from the vast volumes of data generated by IIoT devices, enabling them to anticipate trends, identify patterns, and make informed decisions with confidence. As IIoT continues to evolve and mature, its impact on the industrial landscape will only deepen, driving innovation, and shaping the future of industry for years to come.

Pre - Internet (I)	Internet of CONTENT (II)	Internet of Services (III)	Internet of PEOPLE (IV)	Internet of Things (V)
Human to Human	"www"	Web 2.0	Social Media	Machine to Machine
Fixed & Mobile SMS	Email, Entertainment	E Commerce	Facebook, YouTube Twitter	Identi-fication, Tracking Monitoring
SMART Networks	IT Platforms	Phone Applications	Smart Devices & TAGs	DATA Analysis

Exhibit: 2.1 Evolution of Internet of Things

2.2 Evolution of IIoT Technology

The evolution of IIoT technology can be traced back to the convergence of several key factors, including advancements in communication technologies, the miniaturisation of sensors and actuators, and the proliferation of cloud computing and big data analytics.

Initially, IIoT was primarily focused on machine-to-machine (M2M) communication, where machines could exchange data with each other autonomously. However, with the advent of cloud computing and edge computing, IIoT has evolved to incorporate more sophisticated capabilities, such as real-time data processing, predictive analytics, and machine learning.

Furthermore, the integration of cybersecurity measures has been crucial in ensuring the reliability and security of IIoT systems, especially given the critical nature of industrial operations. Robust protocols and standards have been developed to safeguard against cyber threats and ensure the integrity of data transmitted across IIoT networks.

The Industrial Internet of Things (IIoT) holds immense importance in industrial settings, offering transformative benefits that significantly enhance operational efficiency, productivity, and decision-making processes.

Enhancing Operational Efficiency and Productivity:

IIoT, with its ability to connect machinery, equipment, and systems, stands as a transformative force in enhancing operational efficiency and productivity within industrial processes. By enabling seamless communication and coordination, IIoT minimises manual intervention and streamlines workflows, leading to a higher degree of automation and optimization. This automation not only accelerates production cycles but also reduces waste and operational costs, as processes become more streamlined and resource-efficient. Moreover, IIoT facilitates real-time monitoring of production metrics, providing invaluable insights into performance and productivity. Armed with this data, organisations can make timely adjustments to optimise operations, minimise downtime, and maximise output. This real-time monitoring capability not only enhances operational efficiency but also enables organisations to respond swiftly to changing market demands and production requirements. Overall, IIoT empowers industrial enterprises to achieve new levels of efficiency and productivity, driving competitiveness and growth in an increasingly dynamic and competitive business landscape.

Enabling Predictive Maintenance and Minimising Downtime:

The implementation of Industrial Internet of Things (IIoT) introduces a paradigm shift in maintenance strategies, notably through its capability to enable predictive maintenance. By outfitting industrial assets with sensors and connectivity, IIoT systems revolutionise maintenance practices by continuously monitoring equipment health and performance in real-time. This proactive approach to maintenance allows for the early detection of potential issues and anomalies, enabling maintenance teams to intervene pre-emptively before they escalate into costly failures.

By leveraging predictive analytics and machine learning algorithms, IIoT systems can analyse the vast volumes of data collected from sensors to identify patterns and trends indicative of impending equipment failures. Armed with this insight, maintenance teams can prioritise and schedule maintenance activities more effectively, optimising resource allocation and minimising disruptions to operations.

As a result, unplanned downtime is significantly reduced, leading to enhanced operational efficiency and productivity. Additionally, the early detection and resolution of equipment issues extend the overall reliability and lifespan of industrial assets, thereby maximising their return on investment. This not only results in significant cost savings by eliminating the need for emergency repairs and replacements but also ensures increased uptime and customer satisfaction.

Furthermore, predictive maintenance enabled by IIoT fosters a proactive and data-driven approach to asset management, enabling organisations to transition from reactive to preventive maintenance strategies. By embracing predictive maintenance through IIoT, industrial enterprises can unlock new levels of efficiency, reliability, and competitiveness in an increasingly dynamic and demanding business environment.

Facilitating Data-Driven Decision-Making in Industries:

The Industrial Internet of Things (IIoT) is entering a new era of data-driven decision-making by generating vast amounts of data from sensors, machines, and connected devices within industrial environments. This data serves as a treasure trove of valuable insights into various aspects of industrial operations, ranging from production efficiency and equipment performance to resource utilisation and supply chain management. Through the utilisation of advanced analytics and machine learning algorithms, IIoT systems possess the capability to process and analyse this data in real-time, unveiling patterns, trends, and correlations that may otherwise remain hidden or overlooked by traditional methods. By harnessing the power of real-time data analytics, decision-makers are equipped with actionable insights that empower them to make informed, data-driven decisions aimed at optimising processes, enhancing quality, and fostering innovation. For instance, IIoT data analytics can enable predictive maintenance strategies by identifying early warning signs of equipment failure, thereby minimising downtime and maximising operational efficiency. Additionally, IIoT analytics can optimise production schedules by identifying bottlenecks or inefficiencies in workflows, enabling organisations to allocate resources more effectively and improve overall productivity.

Furthermore, IIoT data analytics can facilitate continuous improvement initiatives by providing feedback loops that enable organisations to iteratively refine and enhance their processes over time. By leveraging IIoT-generated data and advanced analytics, industrial enterprises can unlock new opportunities for innovation and competitive advantage in an increasingly data-driven world. However, it is essential for organisations to invest in robust data management and analytics capabilities, as well as cybersecurity measures, to ensure the integrity, security, and privacy of IIoT data.

Overall, IIoT data analytics holds the potential to revolutionise industrial operations by enabling organisations to harness the power of data to drive efficiency, quality, and innovation across their operations

2.3 Futuristic Nature of IIoT

The Industrial Internet of Things (IIoT) epitomises a futuristic paradigm where interconnected systems and intelligent machines converge to revolutionise industrial operations. At its core, IIoT leverages cutting-edge technologies such as artificial intelligence, machine learning, big data analytics, and edge computing to create a dynamic ecosystem where machines communicate, collaborate, and make autonomous decisions in real-time. This interconnectedness not only enables unprecedented levels of efficiency, productivity, and reliability but also lays the foundation for transformative advancements in industrial automation, predictive maintenance, and supply chain management. Looking ahead, the future of IIoT promises even greater innovation and disruption, with advancements in areas such as 5G connectivity, digital twins, blockchain integration, and quantum computing poised to unlock new possibilities for industrial optimization and innovation. As IIoT continues to evolve, it holds the potential to redefine the very fabric of industrial operations, empowering organisations to thrive in an increasingly digital and interconnected world.

The widespread adoption of the Industrial Internet of Things (IIoT) brings about profound implications for human life, reshaping job roles, enhancing workplace safety, and altering socio-economic dynamics in significant ways. With the integration of IIoT technologies into industrial environments, traditional job roles are undergoing transformation, with a shift towards more data-centric and technology-driven roles.

While some routine tasks may become automated, there is a growing demand for skilled workers capable of managing and interpreting the vast amounts of data generated by IIoT systems. Moreover, IIoT contributes to improved workplace safety by enabling real-time monitoring of equipment and environmental conditions, thereby reducing the risk of accidents and injuries.

By providing timely insights into potential hazards and facilitating proactive maintenance, IIoT enhances worker well-being and productivity. Furthermore, the adoption of IIoT has broader socio-economic implications, fostering innovation, driving efficiency, and creating new opportunities for economic growth and development.

However, it also raises concerns about data privacy, security, and the potential displacement of jobs due to automation. Therefore, as IIoT continues to proliferate, it becomes imperative for policymakers, businesses, and society as a whole to address these challenges proactively and ensure that the benefits of IIoT are realised in a manner that is inclusive, sustainable, and equitable for all.

2.4 Influence on Job Roles and Skill Requirements

The widespread adoption of Industrial Internet of Things (IIoT) precipitates a fundamental shift in job roles and skill requirements across various industries. As automation and connectivity become more prevalent, traditional job functions undergo a transformation, necessitating the acquisition of new skills and competencies. Employees must adapt to working alongside smart machines and leveraging data-driven insights, leading to the emergence of roles such as data scientists, IoT specialists, and automation engineers. Furthermore, there is an increased demand for skills related to cybersecurity, data analytics, and digital literacy to effectively harness the potential of IIoT technologies. This evolution in job roles reflects the ongoing digital transformation and underscores the importance of continuous learning and upskilling in today's dynamic industrial landscape.

For example:

- Technicians and maintenance workers require proficiency in IIoT technologies to perform predictive maintenance and troubleshoot connected machinery.

- Data analysts and engineers are in high demand to analyse vast amounts of sensor data and derive actionable insights for process optimization and decision-making.

- Cybersecurity professionals play a critical role in safeguarding IIoT networks and systems against cyber threats, ensuring the integrity and security of industrial operations.

Moreover, IIoT drives the need for cross-disciplinary skills, such as proficiency in data analytics, programming, and communication, as industries seek to harness the full potential of interconnected technologies.

Improvements in Workplace Safety and Working Conditions:

IIoT contributes to enhanced workplace safety and working conditions through real-time monitoring, predictive analytics, and automation. By deploying sensors and wearables, IIoT systems can detect potential hazards, monitor environmental conditions, and alert workers to dangers in real-time. This proactive approach to safety minimises the risk of accidents and injuries, ensuring a safer working environment for employees.

IIoT enables the implementation of ergonomic enhancements and adaptive workflows that reduce physical strain and fatigue, improving overall worker well-being and productivity. For example, wearable devices can provide real-time feedback on posture and movement, helping workers avoid repetitive strain injuries and musculoskeletal disorders.

2.5 Socio-Economic Implications of Widespread IIoT Adoption

The widespread adoption of IIoT has far-reaching socio-economic implications, affecting industries, economies, and societies at large. Some of the key implications include:

- Increased productivity and efficiency: IIoT enables organisations to optimise processes, reduce waste, and enhance productivity, leading to economic growth and competitiveness.

- Job displacement and re-skilling: While IIoT creates new job opportunities, it also displaces traditional roles through automation and digitization. This necessitates re-skilling and upskilling initiatives to ensure a workforce that is equipped to thrive in the digital age.

- Economic inequality: The benefits of IIoT adoption may not be evenly distributed, leading to disparities in access to technology, skills, and employment opportunities. Addressing these inequalities requires proactive measures to promote inclusivity and equal access to education and training.

- Environmental sustainability: IIoT facilitates resource optimization and energy efficiency, contributing to sustainable development goals and environmental conservation efforts. By minimising waste and emissions, IIoT helps mitigate the impact of industrial activities on the environment, creating a more sustainable future for generations to come.

Key Opportunities and Benefits of IIoT

The Industrial Internet of Things (IIoT) presents a myriad of opportunities and benefits that propel innovation, efficiency, and competitiveness within industrial settings. At its core, IIoT facilitates the seamless integration of physical devices, sensors, and machines with digital technologies, enabling real-time monitoring, analysis, and control of industrial processes. This interconnectedness not only enhances operational visibility but also unlocks new avenues for optimization and automation across various domains, including manufacturing, supply chain management, and asset maintenance.

By harnessing the power of IIoT, organisations can optimise production processes, reduce downtime, and enhance overall productivity. Additionally, IIoT enables predictive maintenance strategies, allowing organisations to anticipate equipment failures before they occur and schedule maintenance proactively, thereby minimising disruptions and optimising asset performance. Moreover, IIoT data analytics provide valuable insights into production metrics, enabling informed decision-making and continuous improvement initiatives. Furthermore, IIoT fosters innovation by enabling the development of new products and services, such as remote monitoring solutions, predictive analytics platforms, and digital twins. Overall, the key opportunities and benefits of IIoT lie in its ability to drive efficiency, agility, and competitiveness, positioning organisations for success in an increasingly digital and interconnected world.

Creating New Business Models and Revenue Streams:

IIoT enables organisations to innovate and create new business models that capitalise on data-driven insights and connectivity. By offering value-added services such as predictive maintenance, remote monitoring, and asset tracking, companies can generate recurring revenue streams and differentiate themselves in the market. Additionally, IIoT facilitates the transition from product-centric to service-centric business models, where companies sell outcomes rather than just products, opening up new revenue opportunities and enhancing customer relationships

Enhancing Product Quality and Customer Satisfaction:

IIoT enables real-time monitoring and control of production processes, ensuring consistent quality and adherence to specifications. By leveraging sensors and analytics, manufacturers can detect defects and deviations early in the production cycle, minimising waste and rework. Furthermore, IIoT enables continuous improvement through feedback loops and data-driven optimization, resulting in products that better meet customer needs and expectations. Ultimately, this leads to higher levels of customer satisfaction and loyalty, strengthening the competitiveness of businesses in the marketplace.

Enabling Remote Monitoring and Control of Industrial Processes:

IIoT allows for remote monitoring and control of industrial processes, providing operators with real-time visibility and insights into operations regardless of their location. This capability is especially valuable for industries with distributed operations or remote facilities, where manual monitoring and intervention may be impractical or costly. With IIoT, operators can remotely monitor equipment performance, adjust process parameters, and respond to anomalies promptly, improving operational efficiency and reducing downtime. Additionally, remote access enables faster troubleshooting and maintenance, minimising disruptions and maximising uptime.

Leveraging Big Data Analytics for Actionable Insights:

IIoT generates vast amounts of data from sensors, devices, and systems throughout the industrial ecosystem. By harnessing big data analytics, organisations can extract actionable insights from this data to drive informed decision-making and optimise operations. IIoT analytics enable predictive maintenance, anomaly detection, and process optimization, allowing companies to anticipate issues before they occur, optimise resource utilisation, and improve overall efficiency. Moreover, analytics empower organisations to identify trends, patterns, and correlations in data, uncovering hidden opportunities for innovation and improvement.

2.6 Challenges and Considerations in IIoT Implementation

Deploying the Industrial Internet of Things (IIoT) presents a host of challenges and considerations that organisations must navigate to achieve successful implementation and widespread adoption. One significant hurdle lies in managing the sheer complexity and scale of IIoT systems, which involve integrating a multitude of devices, sensors, and networks across diverse industrial environments. This complexity can lead to interoperability issues, as different devices and systems may use varying protocols and standards, hindering seamless communication and data exchange.

Additionally, ensuring the security and privacy of data transmitted and stored within IIoT networks is paramount, given the sensitive nature of industrial operations and the potential consequences of cyberattacks or data breaches. Moreover, organisations must grapple with the challenge of effectively managing and analysing the vast volumes of data generated by IIoT devices, as well as implementing robust data governance and management practices to derive actionable insights while maintaining compliance with regulatory requirements.

Addressing Cybersecurity Concerns:

IIoT systems are susceptible to cybersecurity threats, including hacking, data breaches, and malware attacks. Protecting sensitive data and ensuring the integrity of IIoT networks is paramount to safeguarding industrial operations. Organisations must implement robust cybersecurity measures, including encryption, authentication, access control, and intrusion detection systems, to mitigate the risk of cyber threats and ensure the security of IIoT deployments.

Ensuring Interoperability and Compatibility of IIoT Devices:

The diverse and fragmented nature of IIoT ecosystems can lead to compatibility issues between devices, platforms, and protocols. Ensuring interoperability and seamless integration between different components is essential to maximise the value of IIoT deployments. Organisations should adhere to industry standards and protocols, invest in interoperable technologies, and collaborate with vendors and partners to address compatibility challenges and facilitate seamless communication and data exchange.

Overcoming Resistance to Change and Cultural Barriers:

Resistance to change and cultural barriers can hinder the adoption and implementation of IIoT initiatives within organisations. Employees may be reluctant to embrace new technologies or workflows, fearing job displacement or disruption to established processes. Effective change management strategies, including communication, training, and stakeholder engagement, are critical to overcoming resistance and fostering a culture of innovation and collaboration.

2.7 Emerging Technologies Shaping the Future of IIoT

2.7.1 Edge Computing:

Edge computing is poised to revolutionise the future of the Industrial Internet of Things (IIoT) by bringing computation and data storage closer to the point of data generation. Edge computing brings computation and data storage closer to the source of data generation, reducing latency and enabling real-time processing of IIoT data. By performing analytics and decision-making at the edge, IIoT systems can respond rapidly to changing conditions and optimise resource utilisation.

This proximity minimises latency, enabling real-time processing and analysis of IIoT data. By leveraging edge computing, organisations can make faster decisions, improve responsiveness, and reduce reliance on centralised cloud infrastructure. This trend not only enhances operational efficiency but also unlocks new possibilities for innovation and agility in industrial settings, paving the way for a more connected and responsive industrial ecosystem.

2.7.2 Artificial Intelligence (AI) and Machine Learning (ML):

Artificial intelligence (AI) and machine learning (ML) algorithms are indispensable components of the Industrial Internet of Things (IIoT), empowering organisations to extract actionable insights from the vast volumes of data generated by connected devices and sensors. By analysing this data, AI and ML algorithms uncover patterns, trends, and anomalies, enabling predictive maintenance strategies to minimise downtime and optimise asset performance. AI and machine learning algorithms play a pivotal role in IIoT by analysing vast amounts of data to uncover patterns, trends, and anomalies. These insights enable predictive maintenance, optimization of production processes, and autonomous decision-making, driving efficiency and innovation in industrial operations.

Moreover, these technologies facilitate process optimization by identifying inefficiencies and automating decision-making processes, thereby enhancing operational efficiency and driving innovation in industrial operations. The integration of AI and ML in IIoT represents a transformative shift towards more intelligent, adaptive, and responsive industrial ecosystems.

2.7.3 5G Connectivity

The adoption of 5G technology heralds a new era for Industrial Internet of Things (IIoT) applications, offering ultra-fast and low-latency connectivity that unleashes a multitude of possibilities. With 5G, IIoT devices can transmit large volumes of data in real-time, facilitating high-definition video monitoring, remote control of industrial processes, and seamless collaboration across distributed environments. The advent of 5G technology promises ultra-fast and low-latency connectivity, unlocking new possibilities for IIoT applications. With 5G, IIoT devices can transmit large volumes of data in real-time, enabling high-definition video monitoring, remote control of industrial processes, and enhanced collaboration across distributed environments. This enhanced connectivity not only improves the efficiency and responsiveness of industrial operations but also enables the implementation of innovative use cases such as augmented reality maintenance, real-time asset tracking, and immersive training simulations. Ultimately, 5G technology empowers organisations to unlock the full potential of IIoT, driving productivity, efficiency, and competitiveness in industrial settings.

2.8 Anticipated Advancements and Potential Disruptions in Industrial Automation

2.8.1 Robotics and Automation: The integration of Industrial Internet of Things (IIoT) with robotics and automation technologies marks a transformative shift in manufacturing and logistics operations. Collaborative robots (cobots) armed with sensors and connectivity are set to revolutionise industrial environments by working alongside human workers. This collaboration enhances productivity, safety, and flexibility, as robots can perform repetitive or dangerous tasks while humans focus on more complex activities. With IIoT-enabled connectivity, these cobots can communicate with each other and with centralised systems, enabling real-time coordination and optimization of workflows. This integration not only streamlines operations but also fosters a more agile and responsive manufacturing and logistics ecosystem, driving innovation and competitiveness in the industry.

2.8.2 Digital Twins: Digital twin technology creates virtual replicas of physical assets, processes, and systems, enabling real-time monitoring, simulation, and optimization. These digital twins facilitate real-time monitoring, simulation, and optimization, leveraging IIoT data and analytics to enable predictive maintenance, scenario analysis, and virtual testing of new products and processes.

By providing a digital counterpart to physical assets, digital twins empower organisations to identify potential issues before they occur, optimise performance, and accelerate innovation cycles. This results in reduced downtime, increased operational efficiency, and faster time-to-market for new products and processes, driving competitive advantage and fuelling growth in industrial sectors.

2.8.3 Blockchain technology offers secure and transparent data management solutions for IIoT applications. By providing a tamper-proof and decentralised ledger, blockchain ensures data integrity and enables trustless transactions between IIoT devices, suppliers, and stakeholders, fostering transparency and accountability in industrial ecosystems.

2.8.4 Ethical and Regulatory Considerations:

As the adoption of Industrial Internet of Things (IIoT) technologies continues to grow, organisations face a pressing need to navigate ethical and regulatory considerations associated with the collection, storage, and analysis of industrial data. Ensuring data privacy, security, and compliance with regulations such as GDPR and industry-specific standards becomes paramount. Moreover, ethical dilemmas may arise regarding data ownership, consent, and potential biases in algorithms. Organisations must implement robust data governance frameworks, transparent data practices, and ethical guidelines to build trust with stakeholders and mitigate risks. By addressing these ethical and regulatory considerations proactively, organisations can foster responsible and sustainable deployment of IIoT solutions while maximising the benefits of data-driven innovation in industrial settings.

2.8.5 Privacy Implications: Industrial data collection and analysis pose privacy concerns, especially concerning employee surveillance. To address this, organisations need robust policies ensuring ethical data use and safeguarding privacy rights. Such measures are essential for upholding employee trust and complying with data protection regulations. Transparency in data collection practices, obtaining informed consent, and limiting data access are vital aspects. Additionally, implementing anonymization techniques and regular audits can mitigate privacy risks. By prioritising ethical data handling, organisations can foster a culture of trust and accountability, promoting employee well-being and maintaining regulatory compliance in the era of industrial digitalization.

2.8.6 Compliance with Data Protection Regulations: IIoT deployments must comply with data protection regulations and industry standards, such as the General Data Protection Regulation (GDPR) in Europe and the California Consumer Privacy Act (CCPA) in the United States. It's imperative for organisations to enact safeguards for data security, acquire consent for data collection, and maintain transparency regarding data handling practices. This ensures compliance with regulatory frameworks and upholds individuals' privacy rights. Measures such as encryption, access controls, and data anonymization help safeguard sensitive information. Moreover, organisations should provide clear disclosures about data usage and processing to foster trust with stakeholders. By prioritising regulatory compliance and ethical data practices, businesses can mitigate risks and build confidence in their IIoT deployments.

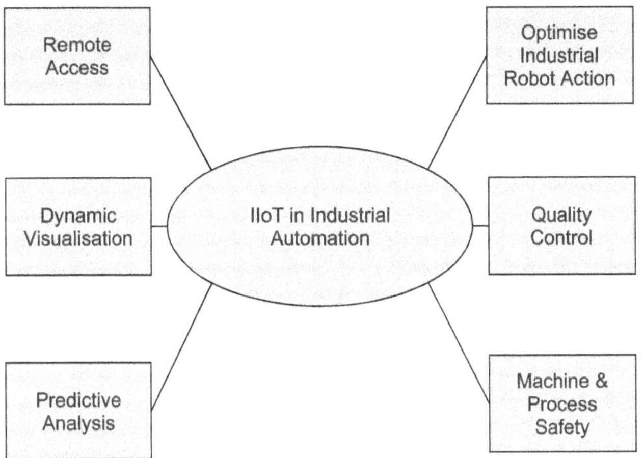

Exhibit 2.2 Significance of IIoT in Industrial Automation

2.9 IIoT applications

Business applications of the Internet of Things (IIoT) are diverse and far-reaching, responding to evolving businesses and processes around the world. In this section, we examine three key areas where IIoT will have a significant impact:

- Smart Manufacturing,

- Supply Chain Management,

- and Energy Management.

2.9.1 Smart Manufacturing

- **Real-Time Monitoring:** One of the main applications of Industrial IoT in manufacturing is real-time monitoring. Sensors embedded in machinery and equipment collect large amounts of data, giving workers and managers unprecedented visibility into the production process.

- **Performance Optimization:** Workers can instantly monitor machine performance, detect bottlenecks, and improve production processes to increase efficiency and reduce downtime.

- **Quality Assurance:** Real-time monitoring allows continuous monitoring of product quality. Any differences in quality can be detected and resolved quickly, ensuring only the best products reach the market.

- **Predictive Maintenance:** The impact of Industrial IoT on maintenance and production linkages. Predictive maintenance, enabled by continuous monitoring of equipment through sensors, is changing the rules of the business world.

- **Reduce downtime:** Predictive maintenance algorithms analyse data from sensors to predict when equipment will fail. This approach minimises unplanned downtime by allowing maintenance teams to address issues before they escalate.

- **Cost Savings:** Organisations can reduce maintenance costs by replacing reactive maintenance with predictive models. Unplanned medical emergencies and related expenses are reduced.

- **Quality Control:** With the use of IIoT, production control has reached a new level.

- **Real-time quality monitoring:** IIoT can continuously monitor products throughout the entire production process. Sensors detect changes and deviations so adjustments can be made immediately to maintain consistent patterns.

- **Traceability:** IIoT facilitates end-to-end traceability by providing detailed information about the manufacturing process of each product. This is important for quality assurance, compliance and, if necessary, remediation.

2.9.2 Supply Chain Management

- **End-to-End Visibility** IIoT provides transparency and visibility to the entire supply chain, from raw materials to purchasing and delivery of finished products.

- **Instant Tracking:** Sensors and RFID technology can instantly locate items throughout the supply chain. This view allows organisations to track the status and location of shipments, reduce the risk of delays and improve overall performance.

- **Demand Forecasting:** IIoT data enables more accurate demand forecasting. Organisations can analyse historical data and current events to make informed predictions, improve product quality, and reduce the risk of product outages or overstocks.

- **Inventory Management:** The Internet of Things is changing traditional inventory management by increasing accuracy and efficiency. Automatic inventory tracking: RFID tags and sensors can track product levels. This automation reduces manual errors, increases accuracy and speeds up the process.

- **Dynamic reordering:** With real-time information on inventory and usage patterns, organisations can use dynamic reordering. This extends the duration of the goods sold, reduces the handling cost and prevents the goods from leaving.

- **Logistics and Transportation:** In logistics and transportation, IIoT optimises operations and increases overall efficiency.

- **Fleet Management:** Vehicles equipped with Industrial IoT provide instant information about their location, status and performance. This leads to more efficient planning, fuel optimization and efficient scheduling.

- **Condition monitoring:** Sensors monitor the condition of cargo during transportation to ensure proper handling of defective or damaged products. This information is important for product quality control and compliance with shipping regulations.

2.9.3 Energy Management

- **Energy Efficiency** IIoT plays an important role in ensuring energy efficiency in production facilities.

- **Energy Monitoring:** Sensors and smart metres collect data on energy consumption. This information allows organisations to analyse their energy efficient processes, implement energy saving measures and increase overall efficiency.

- **Demand Response:** IIoT can respond to changes in energy demand. By adjusting energy consumption according to demand patterns, organisations can participate in demand response programs, reduce costs and support security plans.

- **Environmental Sustainability:** Industrial IoT not only increases efficiency but also promotes environmental sustainability measures in business.

- **Reducing waste:** Industrial IoT helps monitor the production process to identify and reduce waste. Organisations contribute to environmental sustainability by optimising resource use and following responsible practices.

- **Emissions Monitoring:** Sensors can monitor emissions and environmental impacts. This information helps organisations comply with regulations, reduce their carbon footprint and achieve sustainability goals.

Business applications of the Internet of Things have changed the fields of production, supply chain management and energy consumption. From real-time monitoring and predictive maintenance to supply chain optimization and environmental safety, the IoT Industry is driving efficiency, innovation, and responsibility.

Keynotes

Topic	Keynotes
IIoT revolutionizes industrial operations	Through interconnected smart devices, enabling real-time data exchange and automation.
Advancements in IIoT technology	Driven by communication, sensor miniaturization, and cloud computing advancements.
Benefits of IIoT	Operational efficiency, predictive maintenance, data-driven decision-making, improved workplace safety.
Challenges in IIoT implementation	Cybersecurity concerns, interoperability, compatibility, resistance to change, and cultural barriers.
Opportunities presented by IIoT	New business models, revenue streams, enhanced product quality, customer satisfaction, and improved competitiveness.
Real-world applications of IIoT	Span across manufacturing, agriculture, supply chain management, energy management, and development of smart products/services.
Emerging technologies shaping the future of IIoT	Edge computing, AI, ML, 5G connectivity, and blockchain, promising further advancements and disruptions in industrial automation.

Trivia Challenge Industry 4.0

Welcome to the Industry 4.0 Trivia Challenge! Get ready to test your knowledge of the wacky world of Industry 4.0 with these hilarious multiple-choice questions. Each question is paired with some side-splitting facts and figures about Industry 4.0. Let us dive in and have a giggle!

1) What is the Industrial Internet of Things (IIoT) often referred to as?

    A) The Internet of Industrial Stuff (IoIS)

    B) The Internet of Sassy Things (IoST)

    C) The Internet of Tinkering Toys (IoTT)

    D) The Internet of This and That (IoT&T)

2) According to the chapter, how does IIoT impact job roles?

    A) It turns everyone into robot whisperers.

    B) It transforms job titles into something out of a sci-fi movie.

    C) It replaces humans with sentient toasters.

    D) It makes everyone suddenly interested in coding and data analysis.

3) What contribution does IIoT make to workplace safety?

    A) It installs bubble wrap on all the machinery.

    B) It replaces "Caution: Wet Floor" signs with holographic warnings.

    C) It uses sensors and real-time monitoring to prevent accidents.

    D) It hires a squad of safety-conscious robots to patrol the premises.

4) What are the socio-economic implications of widespread IIoT adoption?

    A) People start trading Bitcoin for wrenches on the factory floor.

    B) It creates new opportunities for AI-powered coffee breaks.

    C) It leads to an increase in efficiency and productivity.

    D) It sparks debates on whether robots should get vacation days.

5) How are IIoT technologies described in the chapter?

    A) As magical fairy dust sprinkled over factories.

    B) As the secret sauce that makes everything smarter.

    C) As the behind-the-scenes wizards pulling the strings.

    D) As the Avengers assemble to save the manufacturing world.

6) Which industry is NOT mentioned as having practical IIoT applications?

    A) Chocolate teapot manufacturing (just kidding!).

    B) Healthcare

    C) Agriculture

    D) Fashion (bright sweaters are still a work in progress).

7) **What is the role of emerging technologies in shaping the future trajectory of IIoT?**

    A) They're like the crystal ball predicting when the coffee machine will break down.

    B) They're the rocket fuel propelling IIoT to new heights.

    C) They're the superglue holding together the IoT-connected universe.

    D) They're the wild cards in the deck of the Industrial Revolution.

8) **According to the chapter, how does IIoT improve competitiveness and market positioning?**

    A) By turning factories into hotbeds of techno-wizardry.

    B) By giving companies an edge sharper than a data-driven samurai sword.

    C) By flooding the market with IoT-enabled gizmos and gadgets.

    D) By transforming businesses into digital juggernauts poised for world domination.

9) **What does IIoT stand for?**

    A) Intelligent Internet of Things

    B) Insanely Incredible Over-the-top Technology

    C) Industrial Internet of Things

    D) Incredibly Intriguing Ornamental Trinkets

10) **According to the chapter, in what way does IIoT enhance customer engagement?**

    A) By sending them personalized birthday messages from their favourite machines.

    B) By creating a virtual reality shopping experience inside the factory.

    C) This Will allow them to track their product's journey from the assembly line to the doorstep.

    D) Give them a virtual high-five every time they interact with a smart device.

Summary:

The Industrial Internet of Things (IIoT) has revolutionized industrial operations by creating interconnected ecosystems of smart devices, sensors, machines, and systems that exchange real-time data. Its evolution has been driven by advancements in communication, miniaturization of sensors, and the proliferation of cloud computing and big data analytics. IIoT offers transformative benefits, including operational efficiency, predictive maintenance, data-driven decision-making, and improved workplace safety. However, its widespread adoption also brings challenges such as cybersecurity concerns and resistance to change. Despite these challenges, IIoT presents numerous opportunities for businesses, including new revenue streams, enhanced product quality, customer satisfaction, and improved competitiveness. Real-world applications of IIoT span across manufacturing, agriculture, supply chain management, energy management, and the development of smart products and services. Emerging technologies like edge computing, AI, ML, 5G connectivity, and blockchain continue to shape the future of IIoT, promising further advancements and disruptions in industrial automation.

Abbreviations:

Abbreviation	Terms
IIoT	Industrial Internet of Things
M2M	Machine-to-Machine
AI	Artificial Intelligence
ML	Machine Learning
5G	Fifth Generation
IoT	Internet of Things
ML	Miniaturization
QC	Quality Control
DA	Data Analytics
SM	Smart Manufacturing
SCM	Supply Chain Management
EMS	Energy Management
DCS	Digital Control System
SR	Strategic Positioning
FCT	Futuristic Cities

Chapter 3

Basics of Communication Network: IIoT

Learning Objectives:

- Understand the fundamentals of the Internet of Things (IoT) and its potential applications in various domains.

- Identify the key components and technologies involved in IoT network development.

- Explore the OSI model and its relevance in understanding network fundamentals for IoT.

- Recognize various IoT networking technologies and their applications in different scenarios.

- Analyse the challenges and considerations in IoT networking, including range, bandwidth, power transfer, intermittently connected devices, and interoperability.

- Examine the IoT reference model and its significance in breaking down complex IoT systems into manageable parts.

- Define the Internet of Things (IoT) ecosystem and it essential elements: devices, networks, platforms, and agents.

- Explore specific technologies and standards such as IEEE802.15.4 and ZigBee in the context of industrial wireless sensor networks.

3.1 Fundamentals of IIoT

Today, the internet has become widespread, touching almost every corner of the world and affecting people's lives in unexpected ways. We are now entering an era of greater connectivity, where many devices will be connected to the Internet. What can be said clearly is that the Internet of Things (IoT) is reaching many different people and is becoming increasingly accepted. Potential IoT application areas include smart cities, smart cars and travel, smart homes and living services, smart business, public safety, defence and environment, agriculture and tourism are part of the future IoT ecosystem

There are 2 definitions: First one is defined by Vermesan and second by Pena-Lopez 1. The Internet of Things is simply an interaction between the physical and digital worlds. The digital world interacts with the physical world using a plethora of sensors and actuators.

2. Another is the Internet of Things is defined as a paradigm in which computing and networking capabilities are embedded in any kind of conceivable object.

The Internet of Things represents a new world where almost all devices and equipment we use can connect to the internet. We can use them together to complete complex tasks that require a high level of skill. To facilitate this intelligence and connectivity, IoT devices are equipped with sensors, electronics, processors, and transceivers. The Internet of Things is not a technology; rather, it is a collection of technologies working together. Sensors and actuators are devices that facilitate interaction with the physical environment. Information collected by sensors needs to be stored and processed intelligently so that useful information can be obtained.

Please note that we define the term sensor broadly; A mobile phone or even a microwave oven can be considered sensors as long as it provides feedback about its current state. Actuators are ideal for affecting changes in the environment, such as temperature controls for air conditioning.

Storage and processing of data can be done at the edge of the network or in a remote server. If any action can be taken on the data, it is usually done on a sensor or other nearby smart device.

The processed data is then sent usually to a remote server. The storage capacity and operation of IoT products are still limited by available resources, which are often limited due to size limitations, power, energy, and computing power.

The real challenge of research, therefore, is to ensure that we have the right kind of data at the desired level of accuracy. Along with the challenges of data collection, there are also communication challenges.

Communication between IoT devices is mostly wireless as they are usually installed in separate areas. Wireless channels are often highly sensitive and unreliable. In this case, reliable data communication without excessive retransmission is an important issue, so communication technology is an important part of IoT equipment research.

We can change the state of any device in the physical world from the actuator. For example, we can send messages to other smart devices. The processes that affect change in the physical world often depend on its current state. This is called context awareness. All decisions are made with context in mind, as applications may behave differently in different contexts.

For example, a person may not like to be disturbed by messages from the office while on vacation. Sensors, actuators, computing servers, and communication networks together form the core components of the IoT framework. But there is a lot of software to consider.

First, we need an environment that can be used to connect and manage all these different things. We need more designs to connect different devices.

The Internet of Things has many applications in the fields of medicine, healthcare, education, entertainment, social, energy saving, environmental protection, home automation and transportation.

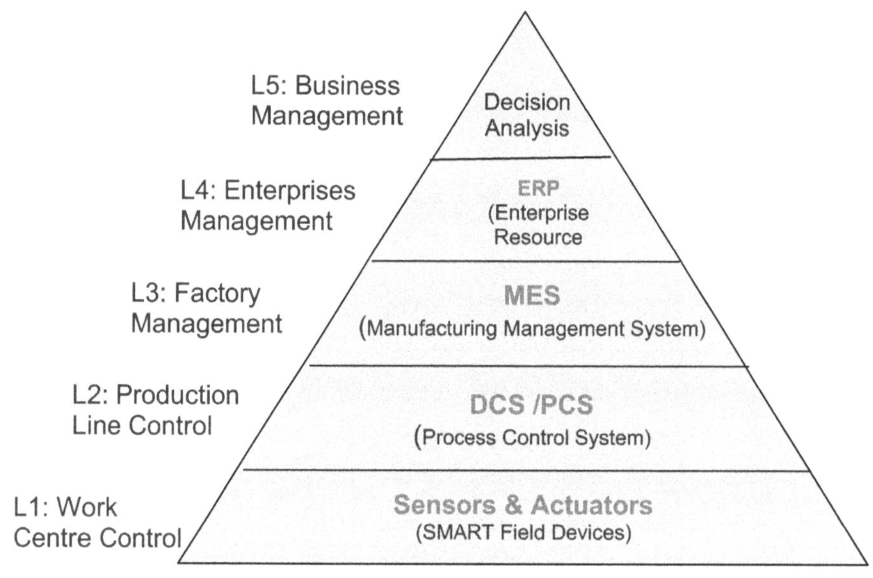

Exhibit 3.1 IIoT Layers

3.2 Internet / WEB and Network Fundamentals OSI Model

Technologies involved in IoT development: Internet and Network Fundamentals OSI Model

Network technologies help IoT devices communicate with other devices, applications and services running in the cloud. The Internet relies on protocols to ensure the security and reliability of different devices. Protocol specifies the rules and standards used by devices used to create, manage networks, and transmit data over those networks.

The Internet was designed as a "stack" of technologies. Technologies like Bluetooth LE are at the bottom of the stack.

In addition to the group, other technologies such as IPv6 technology (responsible for logical addresses and network traffic) are also included. Technologies at the top of the stack are used by applications running above layers, such as message queuing technologies.

IoT is a widely used technology and widely used standards also provide guidance on choosing one network protocol over another. The key considerations and issues related to IoT networking: range, bandwidth, power consumption, connectivity, interoperability, and security.

Network models and technologies

The Open Systems Interconnection (OSI) model is the ISO standard abstract model, which is a group of seven protocol standards The Open Systems Interconnection (OSI) model is the ISO standard abstract model, which is a group of seven protocol standards.

From top to bottom they are: application, presentation, session, transport, network, data link and physical. TCP/IP, or the Internet Protocol suite, underpins the internet, and it provides a simplified concrete implementation of these layers in the OSI model.

Network Access and Physical Layers: This TCP/IP layer consists of OSI layer 1 and layer 2. Which controls how each device connects to the network through hardware such as optical cables, wires, or radio, in the case of a wireless network. At the link layer, devices are identified by their MAC addresses, and operations at this layer include physical aspects such as changing how frames are sent to equipment on the network.

Internet Layer: This process corresponds to OSI Layer 3 (Network Layer). OSI layer 3 is about addressing. This set of layers defines how routers send packets between destinations and addresses defined by IP addresses. IPv6 is mainly used for IoT device addresses.

Transport Layer the Transport Layer (number 4 in OSI) focuses on end-to-end communication, providing reliability, preventing collisions, and ensuring packets are delivered as it is in the same manner. For efficiency, IoT transport typically uses UDP (User Datagram Protocol).

Application Layer The application layer (layers 5, 6, and 7 in the OSI) covers application-level communication. HTTP/S is an example of an application protocol widely used on the Internet. Although the TCP/IP and OSI standards give you a basic framework for discussing network protocols and the specific technologies that use each protocol, some protocols do not conform to these protocols. This layer is not good. For example, the Transport Layer Security (TLS) protocol, which ensures the confidentiality of network connections and data integrity through the use of encryption, can be considered operative at OSI layers 4, 5, and 6.

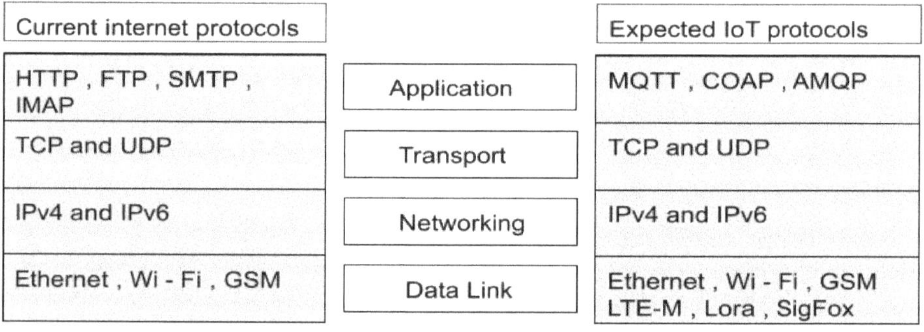

Exhibit 3.2 Internet Protocols and IoT

3.3 Network access and physical layer IoT network technologies

At the bottom of the process stack to understand are IoT network technologies such as cellular, Wifi and Ethernet, as well as various solutions such as -

LPWAN, Bluetooth Low Energy (BLE), ZigBee, NFC and RFID. According to Gartner, NB-IoT has become the standard for LPWAN networks. This IoT for Everyone article covers NB-IoT in detail. The following are networking technologies with a brief description of each technology:

LPWAN (Low Power Wide Area Network) is a type of technology designed for low-power, remote wireless communications. They are ideal for mass deployment of low-power IoT devices such as wireless sensors. LPWAN technologies include LoRa (Long Range Physical Layer Protocol), Haystack, SigFox, LTE-M and NB-IoT (Narrowband Internet of Things).

Cellular LPWAN NB-IoT and LTE-M standards solve the problem of low-power, low-cost IoT communications options using existing cellular networks. NB-IoT is the latest version of this trend and focuses on remote communication between large indoor devices.

LTE-M and NB-IoT were developed specifically for the Internet of Things, but existing cellular technologies are also frequently used for wireless communications. While this includes older technologies such as 2G (GSM) (currently deprecated), CDMA (deprecated or deprecated), it also includes 3G, which is being replaced very quickly as many network providers are deprecating all 3G devices. 4G, 5G are available and will continue to work until completed

Bluetooth Low Energy (BLE) BLE is a low energy Bluetooth 2.4 GHz wireless communication protocol. It is designed for short (no more than 100 metres) communications and is usually in a star configuration with one member controlling various devices. Bluetooth operates at Layer 1 (PHY) and Layer 2 (MAC) of the OSI model. BLE is best for devices that transmit small amounts of data serially. Devices are designed to go into sleep mode and save energy when not sending data. Personal IoT devices such as health wearables and fitness trackers often use BLE.

ZigBee: ZigBee operates in the 2.4 GHz wireless communication spectrum. Its range is up to 100 metres longer than BLE. It also has a lower data rate than BLE (more than 250 kbps compared to BLE's 270 kbps). ZigBee is a mesh networking protocol. Unlike BLE, not all devices can sleep on break. A lot of this depends on where they are in the network and whether they need to act as a router or controller in the network. ZigBee is designed for home use and home automation. Another technology related to ZigBee is Z-Wave, which is also based on IEEE 802.15.4. Z-Wave is designed for home automation. It has always been a heritage property, but has recently been publicly announced.

NFC: The Near Field Communication (NFC) protocol is used for small (up to 4 centimetres) communications, such as placing a tag on an NFC card or card reader. NFC is primarily used for payments but can also be used for identification and smart tags on assets.

RFID: RFID stands for Radio Frequency Identity. RFID tags store codes and information. The tag is attached to the device and read by the RFID reader. RFID range is less than one metre. RFID tags can be active, passive or inactive. Passive tags are ideal for battery-less devices because they are read passively by the ID reader. Active tags broadcast their ID periodically, while passive tags only work when the RFID reader is present. Dash7 is a communication protocol using active RFID designed for secure communication in IoT applications. Similar to NFC, RFID's use cases are to track inventory in stores and IoT applications.

Wi-Fi: Wi-Fi is a wireless networking standard based on the IEEE 802.11a/b/ g/n specification. 802.11n provides the highest data throughput with the most power consumption; hence, IoT devices only use 802.11b or g for energy saving reasons. While Wi-Fi is included in many models and current-generation IoT devices, it looks like Wi-Fi will be replaced by low-power transmissions as longer, lower-resolution solutions become common place.

Ethernet: Widely used in local area network connections, Ethernet uses the IEEE 802.3 standard. Not all IoT devices need wireless. For example, sensor devices installed in building automation systems can use LAN cable networks.

Internet Layer (IoT Network Technologies)

Internet Layer Technologies (OSI Layer 3) analyse and transmit packets. The most common technologies used in IoT are related to this layer, including IPv6, 6LoWPAN, and RPL.

IPv6: At the Internet layer, devices are identified by their IP addresses. IPv6 is mainly used in IoT applications as a replacement for traditional IPv4 addresses. IPv4 is limited to 32-bit addresses and can only provide a total of 4.3 billion addresses; this is less than devices currently connected to IoT, IPv6, on the other hand, uses 128 bits.

IP4	IP6
4.2 Billion unique IP Addresses	3.4 x 1038 unique IP Addresses
32-bit address	128 - bit address
Numeric	Alphanumeric
Bits divided by a period	Bits divided by a colon
Security is dependent on applications	Implements an IP Security (IPSec Protocol)
Has checksum fields	Does not have checksum field
Networks are configured manually or with DHCP	Networks are automatically configured

Exhibit 3.3 IP4 Vs IP6

Not all IoT devices need a public address. Most of the tens of millions of devices expected to be connected to the IoT over the next few years will be deployed in private communications using private addresses and will only communicate with other devices or services on the external network using a gateway.

6LoWPAN: IPv6 The Low Power Wireless Personal Area Network (6LoWPAN) standard allows the use of IPv6 over 802.15.4 wireless networks. 6LoWPAN is mainly used for wireless sensor networks, and the Wire protocol for home appliances also works on 6LoWPAN.

RPL The Internet layer also includes functionality. IPv6 Routing Protocol for Low Power and Lossy Networks (RPL) is designed to run IPv6 over low power networks such as those used by 6LoWPAN. RPL is designed for packets in confined areas, such as wireless sensor networks, where not all devices can be reached and packets will not be lost. RPL can build a picture of the nodes in the network to calculate agreement based on dynamic measurements and constraints (such as reducing energy consumption or latency).

Application layer IoT network technology HTTP and HTTPS are ubiquitous in network applications, and the same is true for the internet of things. RESTful HTTP and HTTPS interfaces are widely distributed. CoAP (Constrained Application Protocol) is similar to heavyweight HTTP and is often used with 6LoWPAN over UDP.

Messaging protocols such as MQTT, AMQP and XMPP are also used in IoT applications: ·

Protocol	RESTful	XMPP	RESTful HTTP	MQTT
Transport	UDP	TCP	TCP	TCP
Messaging	Request/ Response	Publish/ Subscribe Request/ Response	Request/ Response	Public/ Subscribe Request/ Response
2G, 3G, 4G Suitability (1000s nodes)	Excellent	Excellent	Excellent	Excellent
LLN Suitability (1000s nodes)	Excellent	Fair	Fair	Fair
Compute Resources	10KS RAM / Flash	10Ks RAM/ Flash	10KS RAM/ Flash	10KS RAM / FLASH
Success Stories	Utility Field Area Networks	Remote Management of Consumer white goods	Smart Energy Profile 2	Extending enterprise, messaging into an application

Exhibit 3.4 IoT Protocols

MQTT Message Queue Telemetry Transport (MQTT) is a publish/subscribe protocol designed as a messaging protocol. It is used in low bandwidth situations, especially sensors and mobile devices on unreliable networks.

AMQP Advanced Message Queuing Protocol (AMQP) is an open messaging protocol for message-oriented middleware. More specifically AMQP is used by RabbitMQ.

XMPP Extensible Messaging and Presence Protocol (XMPP) was originally designed for human-to-human communication, including instant messaging. This technique is suitable for machine-to-machine (M2M) communication using intermediate objects and XML data.

XMPP is mainly used in smart devices. The technology you choose from this process will depend on the specific needs of your IoT project. For example, for a budget home automation system with lots of sensors, MQTT would be a good choice because, as a simple rule, it's best for sending messages between devices that don't have a lot of storage space or power. and is worn to be used.

3.4 IoT Networking Consideration and challenges

Many networks can be identified by the distance between IoT devices connected to the network for data transfer:

PAN (Personal Area Network) PANs are short-range networks that measure distance in metres, such as fitness equipment., Communicate with mobile apps via BLE.

LAN (Local Area Network) LAN is short to medium distance up to hundreds of metres, such as sensors installed in home automation or factory production lines communicating via WIFI with door equipment installed in the same building.

MAN (Metropolitan Area Network) MAN is long (city-wide), measures distances of up to several kilometres and connects to a network topology, like smart parking sensors installed throughout the city.

WAN (Wide Area Network) WAN is a long distance-measured device such as agricultural equipment installed on large farms or ranches to monitor the microclimate environment across the device. Your network needs to collect data from IoT devices and send it to the desired destination.

Range: Select a network protocol that suits the needs. For example, don't choose BLE for WAN applications running more than a few miles. If moving data over the desired range is difficult, consider edge computing. Edge calculates data directly from the device, rather than from remote locations or elsewhere.

Bandwidth: Bandwidth is the amount of data that can be transmitted at one time. It limits the amount of data that can be collected and sent upstream by IoT devices. Bandwidth is affected by many factors, including the amount of data each device collects and transmits, the number of devices used, whether data is transmitted continuously or in parallel, and whether there are significant peaks, the number of packets on the network, and more. The size of the protocol should generally depend on the amount of data sent. Sending packets containing empty data is useless. Conversely, splitting large chunks of data into many smaller packets creates overhead. Data transfer speeds are not always consistent (i.e. upload speeds may be slower than download speeds). Therefore, if there is bi-directional communication between devices, data transfer needs to be taken into account.

Since wireless and cellular networks always have low bandwidth, consider whether wireless technology is a good choice for potential applications. Consider how all old items should be shipped. One solution is to store less data by reducing the sampling frequency. Therefore, you will catch fewer differences and be able to filter out items on your device and remove irrelevant items. You can reduce data transfer if you write the data before converting it.

However, this process affects the flexibility and analytics. Addition and explosion are not always suitable for time-sensitive or delay-sensitive objects. All these technologies increase the data processing and storage capacity of IoT devices.

Power transfer: Transfer data from your device. Sending data over long distances requires more energy than sending data over short distances. You should take into account the power of the device (such as battery, solar cell or capacitor) and its lifespan. A long and stable lifetime not only ensures greater reliability but also reduces operating costs. There are steps to help you achieve longer energy efficiency. For example, you can put your device into sleep mode when it is idle to extend battery life. Another best practice is to model the power consumption of devices across different products and states in the network to ensure that the device's power and capacity are maintained. It actually matches the power required to transmit the necessary information using network technology.

Intermittently connected IoT devices are not always connected. In some cases, devices are designed to connect over time. But sometimes an unreliable network can cause the device to disconnect due to connectivity issues.

Sometimes quality of service issues can arise, such as using spectrum sharing to eliminate interference or contention between wireless networks. Designs that involve networking and finding solutions to provide seamless service should be a priority in the IoT design environment.

Interoperability Devices work and interact with other devices, equipment, systems and technologies. With so many different devices connected to the IoT, interoperability can be difficult. The use of protocols has been a traditional way of managing interactions on the Internet. The model is approved by the participants of the work, avoiding various designs and orientations. With appropriate standards and stakeholders' acceptance of these standards, conflicting issues and thus coordination problems can be avoided. However, with IoT, standard systems sometimes have difficulty keeping up with innovation and change. These are written and published according to new standards that are currently subject to change and will be released soon. Consider ecosystem technologies.

Are they widely adopted?

Are they opening versus proprietary?

How many implementations are available?

Using these questions to plan your IoT networks help plan better interoperability for a more robust IoT network.

Security is the most important thing. It is critical to choose network technologies that provide end-to-end security, including authentication, access, and open port protection. IEEE 802.15.4 contains security standards that provide security features such as access control, message integrity, confidentiality, and replay protection used by standard technologies such as ZigBee.

Consider the following when creating a secure and reliable IoT network:

Authentication Use security techniques to support authentication of devices, gateways, contact applications, services, and applications. Consider using the X.509 standard for device authentication.

Encryption If you are using WIFI, please use Wireless Protected Access 2 (WPA2) for wireless network encryption. You can also use the Private Pre-Shared Key (PPSK) method. To ensure confidentiality and data integrity of communications between applications, be sure to use TLS or Data Transport Layer Security (DTLS) for unstable connections that are based on TLS but operate over UDP. TLS encrypted application data and keeps it secure.

Port Guard ensures that only ports required for communication with the gateway or domain or service remain open for outside side connections.

All other ports must be blocked or protected by the firewall. When using the Universal Plug and Play (UPnP) feature, the device's ports will be exposed. Therefore, UPnP needs to be disabled on the router.

Internet of Things World Forum (IoT WF) Standardised Architecture In 2014, the IoTWF Architecture Committee published a seven-layer IoT architecture reference model. While many IoT application models exist, the models recommended by the IoT World Forum provide a clear and simple view of IoT, including edge computing, data storage, and access. It provides an easy way to see the Internet of Things through visualisation.

3.5 IoT Reference Model

Each of the seven layers is dedicated to specific functions, and security covers the entire model.

The IoT application model describes the control process from the source (which could be a cloud service or a specific database) to the level edge, including sensors, devices, machines, and others. smart end node types. In general, information starts from the edges of the cluster and moves north towards the centre. Using this model, we can achieve the following goals

Break down the IoT problem into smaller parts

Introduce the different technologies of each layer and how they interact

Define the system in which different products can be provided by different vendors

Leading to interoperability It has the process of defining the interface.

Defines a standard security system managed at the transition between layers of elements beneath each of the seven layers of the IoT standard.

Layer	Internet Protocols	SMART Objects Protocols
Application Layer	HTTP/FTP/SMTP/...	CoAP
Transport Layer	TCP/UDP	UDP
Network Layer	IP4/IP6	6LoWPAN
Link Layer	IEEE 802.3 - Ethernet; IEEE 802.11 - Wireless Network	IEEE802.I5.4e -Local and Metro Network

Exhibit 3.5 Layers of Protocols

Layer 1: Physical Devices and Control Systems The first layer of the IoT infrastructure is physical devices and control systems. This layer is where the "things" in the Internet of Things live, including various end devices and devices that send and receive messages. These "things" range in size from almost microscopic sensors to large machines in factories. The main function is to create information that can be queried and/or managed on the internet.

Layer 2: Connection Layer In the second layer of IoT applications, the focus is on connection. The most important feature of this IoT system is the reliable and timely delivery of data. Specifically, this includes the transmission of Layer 1 equipment and the network, as well as the transmission of the network and data processing occurring at Layer 3 (layer by layer). You can see that the network has all the elements of IoT and is no different from the last mile network

Layer 3: Edge Computing Layer Edge computing is the responsibility of Layer 3. Edge computing is often referred to as the "fog" layer and is discussed in the "Fog Computing" section later in this chapter. Among these processes, the focus is on data reduction and conversion of network data into data that can be stored and processed by higher-level processes. One of the principles of this model is to start data processing as quickly as possible and as close to the edge of the network as possible.

Another important task that takes place at Layer 3 is evaluating the data to see if it can be transferred to higher layers. This also allows components to be replaced or specified, facilitating further processing by other systems. Therefore, the main task is to evaluate the data to see whether the threshold has been exceeded and whether any action or alert needs to be sent.

Upper layer: Layer 4 to Layer 7 The upper layer processes the IoT data generated by the lower layer.

3.6 Internet of Things Ecosystem

The Internet of Things does not exist in a vacuum. Currently a sensor doesn't actually do anything and a group of them unless they are all connected to each other and to a platform that produces data for later use. This is what we call the Internet of Things (IoT) ecosystem, a vast network of interconnected and integrated devices and technologies that professionals use for brand-specific goals like building smart cities. Frankly, IoT has unlimited uses, so we can talk about unlimited IoT ecosystems. But if you boil it down to what's happening in the ecosystem, you get a simple model: Devices collect data and send it over the network to a platform that collects information for agents to use in the future. So we have the main elements of the IoT ecosystem: devices, networks, platforms and agents. Let's talk about them in more detail.

The Internet of Things system has four building blocks: sensors, processors, gateways and applications. To create an effective IoT system, each node must have its own characteristics.

Sensors:

They form the front end of IoT devices. These are called "things". Their main purpose is to collect information from the environment (sensors) or provide information to the environment (actuators).

These should be unique devices that can be identified by their unique IP addresses so that they can be easily identified in large networks.

These must be functional, that is, they must be able to collect real-time data. These can be operated by themselves (autonomous in nature) or by users according to their needs (user-controlled).

Sensor examples include oil sensors, water quality sensors, humidity sensors, etc. takes place.

Processor:

The processor is the brain of the IoT system. Their main task is to process and process the data captured by the sensors to extract important information from so much collected raw data. As a result, we can say that it provides intelligence for informational purposes.

The processor works at the fastest speed and can be easily controlled by the application. They are also responsible for preventing data from being encrypted and decrypted.

Embedded hardware devices, microcontrollers, etc. They are data processing devices because they have a processor on them.

Gateway:

The gateway is responsible for shipping the finished product and sending it to the correct location for use.

So, we can say that the gateway supports back and forth communication of data. Provides network connectivity for data. Network connectivity is essential for communication in any IoT system. LAN, WAN, PAN etc. all are examples of gateways.

Applications:

Applications form the other end of the IoT system. The application is required for the use of all collected data. These cloud applications are responsible for providing meaningful information to the collected data. The application is controlled by the user and is the distribution point for certain services.

Application examples include home automation, security systems, business management, etc. takes place. IoT Devices As we said before, there are many scenarios in which IoT can be adopted, and they all require different devices. Here, at the most basic level, we can talk about sensors (i.e. devices that know things like temperature, movement, objects, etc.) and actuators (i.e. mechanical components such as switches or rotors). But smart solutions rarely use any IoT sensors or actuators. For example, if you are considering a smart robotic surgery, it will need hundreds or even thousands of things to be able to measure the difference and act accordingly. But even seemingly uncomplicated solutions are not actually that simple. Think smart management; for plants to grow, it is not a matter of measuring moisture but also fertility; to the sun etc. Providing sufficient water is also a problem. So you don't need just one, but many sensors and actuators that need to work together

When talking about devices that are important to the IoT ecosystem, IoT gateways is one of them. They are a device that acts as a link between the two, "interpreting" and facilitating easy communication between devices. This brings us to the next part of the puzzle.

Based on what you've read before, you may be thinking: "So, if an automatic door detects my presence and opens itself, is that the Internet of Things?" The door has sensors and actuators, so it's not visible, but this is the Internet of Things. It's not really about anything else. As the name suggests, the Internet of Things requires both objects and the Internet (although there are data transfers that do not use the Internet Protocol).

It can be said that the real power of this concept is in the connection. There are many types of IoT connections, again depending on your needs, from "classics" like Wi-Fi or Bluetooth to more specific and technological devices like Low Power Wide Area Networks (LPWAN). They all have different data transfer rates and speeds, making them more or less suitable for certain deployments. For example, think about smart cars that need high speed and distance data and connect to the smart farm we talked about, but they don't need to be there. IoT Platform Whether in the cloud or not, the IoT platform is always the connector of any IoT ecosystem.

They are the silent leaders responsible for managing the life of the device, so you don't have to worry about them. They are also places where information is collected and recorded so that you can understand it. Due to the variety of platforms available on the market and the diverse requirements of vendors, choosing the "ideal" IoT platform for deployment is the most important but also the most difficult. This should not be taken lightly because it determines whether the IoT ecosystem will work or not.

The right IoT device management platform should be versatile and flexible because the IoT world is so fragmented and ever-changing, and you don't want the core elements of the ecosystem to change. It turned into shock. block for distribution. Additionally, your ecosystem needs to be scalable so it can grow naturally, and it needs to be secure so it can do so without threats. Agents are all people whose actions impact the IoT ecosystem. They may be architects or business owners building IoT installations and platforms. But most importantly, it will be the participants who will ultimately get the results. These complex ecosystems were created for one reason: to be productive and improve the quality of life. Agents decide how devices, networks, and platforms will be used to achieve these results. This is where technology and business come together because business goals often form the IoT ecosystem.

People are an important part of this equation. Ecosystems are created by us, managed by us, and ultimately it is our responsibility to realise their potential. It is the tools that collect the data, but it is the people who understand and use the data. The same goes for networks and platforms, they are necessities of ecosystems but have little value if not for the people who create and adapt to their needs. As mentioned earlier, the IoT ecosystem is a very complex concept that defies simple classification due to its different characteristics, from deployment to distribution. Just like our world, the IoT world has many different ecosystems that are constantly evolving and adapting. What they all have in common are ideas and the people who use them: manufacturers, service providers, developers and businesses. But there are still many differences in this changing landscape, and the technology represented by devices, networks, and platforms is always getting better. Nothing affects this environment more than stagnation and shutdown; Therefore, you should always find new, better tools to help you improve.

3.6.1 IEEE 802.15.4: This is an old but simple wireless protocol used to connect smart devices.

IEEE 802.15.4g and IEEE 802.15.4e: These are development results of 802.15.4, focusing mainly on energy use and distribution in smart cities. IEEE 1901.2a: A mechanism used to connect smart devices over wireless networks. IEEE 802.11ah: This is a technology based on the well-known 802.11 Wi-Fi standard for smart devices.

LoRaWAN: This is a scalable system designed for long distances and low power requirements in licence-free areas.

NB-IoT and other LTE types: This technology is generally preferred by mobile service providers who want to connect to smart devices during the licence period.

Standardizations and consortiums: set of guidelines that govern the regulatory process

Physical layer: wired or wireless paths and frequencies

MAC Layer: Physical layer data link Determining the Media Access Control (MAC) layer that connects the control to the control

Competing technologies: IEEE 802.15.4 IEEE 802.15.4 is a low-cost, wireless access technology. In addition to being low cost and providing the necessary batteries, these devices can be easily installed using a compact system while being flexible and versatile. Various communication networks (including decision-making networks) and data use this technology to solve various IoT applications in the consumer and business sectors. IEEE 802.15.4 is commonly used in the following types of deployments:

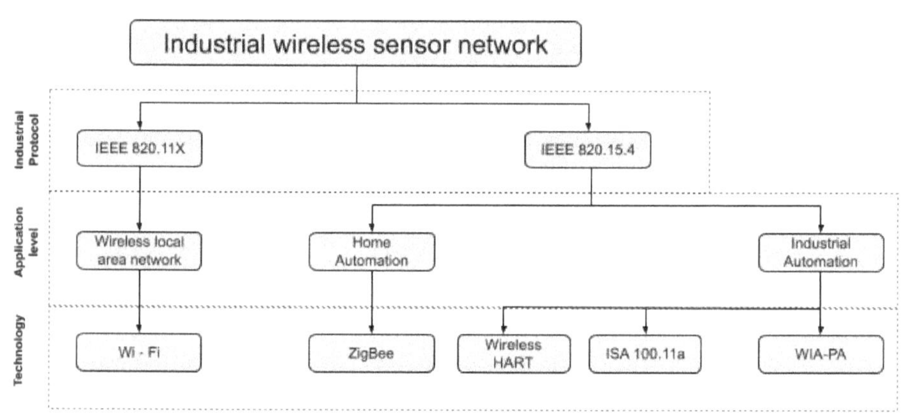

Exhibit 3.6 Industrial Wireless Sensor Networks

3.6.2 Industrial wireless sensor networks

Interactive toys and remote-control IEEE 802.15. The development has mainly focused on MAC reliability, infinite delay, interference and multipath attenuation.

Reliability and latency issues are often associated with the Conflict-Aware Multiple Access/Collision Avoidance (CSMA/CA) algorithm. CSMA/CA is an access method in which a device "listens" to make sure no other equipment is transmitting before initiating its own transmission. If another device is transmitting, there will be a waiting period (usually random) before "listening" again.

Since IEEE 802.15.4 does not have frequency hopping technology, it will be affected and weakened in many ways. IEEE's later version of 802.15.4 began to address these problems.

Standardizations and Associations: IEEE 802.15.4 or IEEE 802.15 Task Group 4 defines low data rate PHY and MAC layer specifications for Wireless Personal Area Networks (WPAN).

The model has been in development for many years and is a good solution for low-cost wireless devices that have low data rates and need months or years of battery life. Since 2003 IEEE has published various aspects of the IEEE 802.15.4 specification. New editions often replace older editions, include appendices, and add functionality or clarification to previous editions. The IEEE 802.15.4 PHY and MAC layers form the basis for stacking various network protocols. This protocol uses 802.15.4 on the physical and outer layers, but the outer layers are different. This process has been developed individually by many organisations and most businesses. The article describes some of the best practices based on the 802.15.4 protocol.

ZigBee: ZigBee is one of the most famous protocols listed in the table. In addition, ZigBee continues to evolve over time, as demonstrated by the release of ZigBee IP, which demonstrates how IEEE 802.15.4 can leverage PHY and MAC protocols independently of the above-mentioned standard procedures. ZigBee Alliance is a trade association that recognizes vendor collaboration and is dedicated to promoting and advancing ZigBee as an IoT solution for connecting smart devices. ZigBee solutions target smart devices and sensors with low bandwidth and low power consumption. Additionally, products that follow the ZigBee standard and are certified by the ZigBee Alliance must be interoperable even if different vendors produce them.

The ZigBee specification has gone through many revisions. The main areas for which ZigBee is best known include automation and smart energy in business, retail and home applications. In business and industrial automation, ZigBee-based devices can perform many tasks, from measuring temperature and humidity to asset tracking. For home automation, ZigBee can control lighting, temperature and security features. ZigBee Smart energy brings together many interconnected devices, such as smart metres, that monitor and control the use and distribution of energy resources such as electricity and water. ZigBee devices managed by energy providers help manage usage between energy providers as well as homes and businesses to ensure greater efficiency. As mentioned before, ZigBee uses the IEEE 802.15.4 standard in the lower PHY and MAC layers. ZigBee defines a network protocol and a security and application support layer on top of these layers.

The ZigBee network and security layer provides a framework for network initialization, configuration, forwarding, and communications protection. This includes calculating the number of trips in transit frequency, finding neighbours, and maintaining meetings when the device first joins.

The network layer is also responsible for creating the required topology (usually a network), but it can be a star or a tree. From a security perspective, ZigBee uses 802.15.4 to provide security at the MAC layer, uses Advanced Encryption Standard (AES) with 128-bit keys, and provides security at the network and application layers.

The support system connects to the lower part of the stack that manages the network of ZigBee devices for higher-level applications. ZigBee comes with a number of pre-built custom applications for specific industries, and vendors can choose to build their own custom applications at this layer.

ZigBee is one of the best known according to IEEE 802.15.4. ZigBee offers its own network and security layers and application profiles, in addition to the 802.15.4 PHY and MAC layers. While this standard provides multi-interoperability between ZigBee Alliance member vendors, it does not provide interoperability with other IoT solutions. However, with the introduction of ZigBee IP, this began to change with the introduction of ZigBee IP, support for IEEE 802.15.4 continues, but the network now supports IP and TCP/UDP protocols, among many other exploits and transport layers.

ZigBee-specific protocols are currently only at the top of the application process stack. It is designed to host open standards from the IETF project at LLN, such as ZigBee IP, IPv6, 6LoWPAN, and RPL. They provide bandwidth-free, low-power and efficient communication when connecting to smart devices. ZigBee IP is a key part of the ZigBee Alliance Smart Energy (SE) Profile 2.0 specification. SE 2.0 targets smart metering and home energy management.

Actually, designed specifically for ZigBee IP SE 2.0 but not limited to this application. Other applications that require protocols based on IoT stacks can use ZigBee IP.

Unlike traditional ZigBee, ZigBee IP supports 6LoWPAN as the switch layer. Since ZigBee IP uses the network or transport path to send packets, the 6LoWPAN network address header is not required. ZigBee IP needs support for 6LoWPAN segmentation and header compression scheme. At the network layer, all ZigBee IP nodes support IPv6, ICMPv6, and 6LoWPAN Neighbour Discovery (ND) and use RPL to forward packets in the mesh network.

TCP and UDP are also supported to provide connection-oriented and connectionless connections.

As you can see, ZigBee IP is a standard manufacturing process because it is based on existing IoT standards at all layers below the application layer. This provides ZigBee IP the opportunity to integrate and collaborate with other solutions developed by the open standard IoT on any 802.15.4 network.

3.6.3 Segmentation

IPv6 networks must have a maximum transmission unit (MTU) of at least 1280 bytes.

Terminology MTU devices have very limited resources, are rarely able to communicate enough to send a few bytes, and have limited security and control; this leads to the need for a standard IP address, where communication occurs through gateways and intermediaries.

A fully functional device and the ability to use separate IP stacking or non-IP stacking: In this case, it is better to use IP stacking and communicate directly with the application server (control model), or use IP or non-IP stacking and a gateway proxy (standard) to communicate replace).

Commonly used in computers and electronic devices, but PC-like devices with limited network resources (such as bandwidth): most nodes use entire separate IP groups (standard), but network design and usage behaviour can govern bandwidth limitations. The definition of constrained nodes continues to evolve. The overall cost of computing power, memory, storage, and power consumption is reduced.

At the same time, technology continues to improve and provide greater bandwidth and reliability. As the use of technology and the cost of many challenges decrease in the future, the effort to develop intellectual property to solve the problem will also decrease. Limited Networks In the early days of the Internet, network bandwidth capacity was limited.

Connections often rely on limited bandwidth for data transmission. However, these low-speed connections show that IP can operate on low-bandwidth networks. But as the Internet grew, so did high-speed performance. However, some IoT devices cannot receive high-speed connections at the last mile. Reasons for this include the use of limited bandwidth equipment, limited bandwidth due to regulatory restrictions, and no more limited Internet service. When the connection process we receive is not available, the network is restricted.

A limited network can cause high latency and packet loss. There are no exceptions or rules to the restrictions. Compared to traditional IP connections, which provide stable and fast connections, wireless connections are limited to low-speed, low-bandwidth connections (wireless and wired). They operate at speeds ranging from a few kbps to hundreds of kbps and can operate efficiently using star, mesh, or converged network topologies. In addition to limited bandwidth, it is normal for the Packet Delivery Rate (PDR) to vary between low and high percentages if there is a conflict in the network. Sometimes many unexpected things can happen and connections can be broken.

These behaviours can be seen in wireless and narrowband networks where packets can change significantly over the course of a day. Unstable peripheral links cause delays and additional problems in aircraft control. One of the golden rules of communication is "There is no excuse for failure." Excessive interference due to low bandwidth and network limitations can disrupt the network and cause existing problems. Flight control should be kept to a minimum, otherwise it will use the bandwidth required to transmit the data. Finally, the energy consumption of the battery from which the voltage is drawn must also be taken into account. Any interference with the flight controls or prolonged use will cause the battery to short circuit.

In summary, node and network limitations cause major problems in connecting IoT to the last mile. This has led many organisations to model their efforts to develop IoT protocols. IP Versions for over 20 years, the IETF has been working to transition the Internet from IP Version 4 to IP Version 6. As the Internet grows, the shortage of IPv4 addresses becomes critical.

IPv6 has many addresses and will not run out in the future. Two types of IP operate on the Internet today, but most traffic is still based on IPv4. While IPv6 may seem like the foundation of any IoT deployment, you need to consider your existing infrastructure and its solutions, processes, and products. IPv4 is so ingrained in today's infrastructure that it needs to be supported in most cases. Therefore, IoT should follow the same path as the web itself by supporting IPv4 and IPv6 versions.

IoT solutions require technologies such as tunnelling and switching to ensure interoperability between IPv4 and IPv6. There are many factors that determine whether IPv4, IPv6, or both can be used in an IoT solution. These situations often involve older devices or technologies that only support IPv4. New technologies and systems almost always support both versions of IP.

Here are some key uses of IPv4 and IPv6 support in IoT solutions:

Functionality: Provide a preferred communication protocol for IoT devices using Ethernet or Wi-Fi interfaces Communication between IPv4 and IPv6 is possible, but IPv4 and IPv6 Communication between is possible, but whether the application protocol is IP will determine the version selection. For example, SCADA protocols such as DNP3/IP (IEEE 1815), Modbus TCP or the IEC 60870-5-104 standard are specified for IPv4 only. So the providers of this protocol no longer use IPv6. Both IP versions are supported by IoT devices using IETF-defined application protocols such as HTTP/HTTPS, CoAP, MQTT, and XMPP. The choice of IP version depends on the application.

Cellular Provider and Technology: IoT devices with cellular modems depend on the type of cellular equipment and information provided by the provider. IPv4 is the underlying protocol of the first three generations of data services (GPRS, Edge and 3G). Therefore, if IPv6 is used in this generation, it must be tunnelled over IPv4. In 4G/LTE networks, data services may use IPv4 or IPv6 as the underlying protocol, depending on the provider. Defines the largest data protocol size that can be sent. For IEEE 802.15.4 the MTU is 127 bytes. This is a problem because IPv6 with large MTUs is carried in 802.15.4 frames with small MTUs. To solve this problem, large IPv6 packets must be split into multiple 802.15.4 frames at Layer 2. The fragmentation header used by 6LoWPAN has three elements: datagram size, datagram flag, and datagram offset. The 1-byte data packet size field shows the total size of the payload error. The datagram tag identifies parts of the payload. Finally, the packet offset field describes how far into the payload a burst occurs.

The 6LoWPAN fragmentation header field itself uses a bit value to identify that the field after it is a fragmentation field and not some other property (such as header compression). Also in the first section the infopack offset field is not available because it is only set to 0. This makes the first part of the IPv6 payload's header only 4 bytes long. The remaining space includes a 5-byte header field to specify the appropriate offset. Mesh Address The purpose of 6LoWPAN's mesh address function is to send packets over multiple hops.

Similar to the IPv6 hop limit field, the network address hop limit provides an upper limit for the number of transmission times. This value is decremented by 1 for each hop during routing. When the value reaches 0, it is discarded and is no longer sent.

The address and address field of the mesh address is the IEEE 802.15.4 address that indicates the end of the IP hop.

Note that mesh address headers are used in an IP subnet with a type of layer 2 routing called mesh-in. RFC 4944 only indicates operation in this case because the meaning of the Layer 2 network routing specifications is beyond the scope of the 6LoWPAN working group and "Layer 2 routing" is not defined by the IETF. 3 IP routing does not need to use network address headers unless the technology requires it.

Application Transfer Method SCADA Internet of Things is a new phenomenon in the use of network technologies and protocols. Coupled with the fact that IP is standard for public computing, legacy methods of connecting sensors and actuators have been modified and better adapted themselves to IP. The most important example of this change is supervisory control and data acquisition (SCADA). SCADA is a control system that has been developed for many years and was initially used without IP cross-connection, later converted to Ethernet and IPv4. Some information about SCADA Over the years, vertical industries have developed communications systems to suit their unique needs. Many of these have been considered and implemented in cases where the most common communication technologies, such as RS-232 and RS-485, are based on serial communication. This leads to the emergence of the SCADA network protocol, which is a good standard compared to other regulations and operates directly on physical systems and data links.

At an advanced level, SCADA systems can collect sensor data and telemetry from remote controls while also controlling them. SCADA systems are used in modern networks to enable global, real-time, data-driven decisions on how to improve business processes. SCADA networks exist in many industries, but you'll find SCADA most focused on the utilities and manufacturing/business sectors. In these specific industries, SCADA often uses some techniques to communicate between equipment and applications. For example, Modbus and its variants are industrial systems used to monitor and generate remote devices through a master/slave relationship. Modbus is also used in building management, transportation and utilities. DNP3 and International Electro Technical Commission (IEC) 60870-5-101 standards, along with DLMS/ COSEM and ANSI C12 for Advanced Reading (AMR), are widely used in the electronics industry. As mentioned earlier, these contracts go back decades and are series based. Therefore, sending them over existing IoT and traditional connections requires some changes in terms of processes and agreements. These services and other changes lead to many SCADA transfers. Adapting SCADA to IP the rapid adoption of Ethernet in business in the 1990s spurred the development of SCADA application protocols. For example, the IEC adopted the Open Systems Interconnect (OSI) layer standard to define its agreement.

Other protocol groups have also slightly adapted their protocols to run on IP infrastructure. Benefits of moving to Ethernet and IP include the ability to support existing devices and models while integrating SCADA subnets into the enterprise WAN infrastructure. To further support legacy business processes of IP networks, special systems have been developed and IP usage information has been published for each system. This involves assigning a TCP/UDP port number to the protocol as follows: via port 20000

DNP3 intelligence

Modbus messaging service using TCP port 502. IEC 60870-5-104 is a modification of IEC 60870. The -5-101 standard is designed to operate over Ethernet and IPv4 using port 2404.

DLMS User Manual DLMS/COSEM or TCP/IP-based data communication in IEC 62056-53 and IEC 62056-47 standards allows data transfer between IP and port 4059 Many other Like the SCADA protocol, DNP3 is based on the master/slave relationship.

In this case, the word "master" refers to the powerful computer in the power control centre, and the word "slave" refers to the remote device where the device is located, including places such as parking.

DNP3 stands for slave station operations. Substations collect data from devices that indicate the generator's status, such as whether it is on or off, and provide information such as voltage, current, temperature, etc. and make measurements. These objects are then sent to the base station when requested, or events and notifications can be sent asynchronously. The master controller also issues control commands, such as starting a motor or resetting a circuit breaker, and records data input. The IEEE 1815-2012 specification describes how the DNP3 protocol should be adapted to TCP (recommended) or UDP. This specification defines the connection management between the DNP3 protocol and the IP layer.

In addition to the configuration and procedures required to use the network connection, connection management also connects the DNP3 layer to the IP layer. The IP layer is transparent to the DNP3 layer because each piece of the protocol at one station logically communicates with the corresponding piece at another station. This means that the DNP3 endpoint or the device is not aware of the IP change that has occurred.

The host initiates the connection by completing a TCP open. The station listens for communication requests by completing TCP passive openings. Finally, the two-point definition is a process that can both listen to communication requests and operate in an open channel when necessary. A base station can identify multiple DNP3 data connections from a single UDP data packet, but DNP3 data connections cannot pass multiple UDP data packets. When the TCP keep alive timer monitors the connection status, one or more connections will be established to the host. The Keep alive message is based on the DNP3 data link request layer. If no response is received to the stored message, the connection is considered broken and necessary actions are taken.

IoT Application Layer Protocols Considering the large-scale deployment of limited and/or limited resources, long-term communication protocols and data structures can pose a challenge for IoT applications. To solve this problem, the IoT industry is looking for new competing methods that accommodate the many limitations of nodes and networks. The two most popular protocols are CoAP and MQTT.

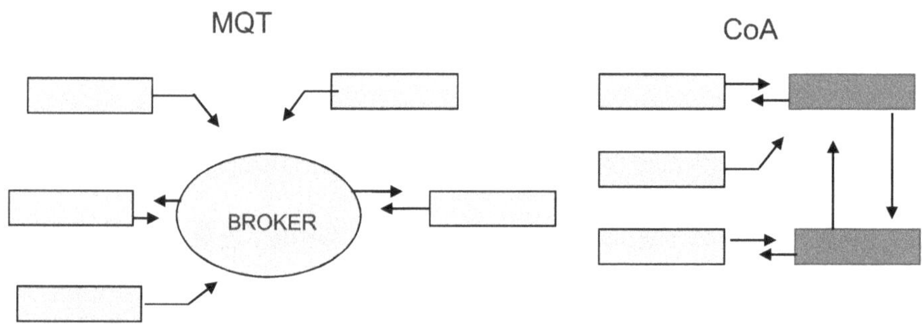

Exhibit 3.7 MQTT Vs CoAP

CoAP and MQTT are always at the top of this example IoT stack based on IEEE 802.15.4 networking. While there are some exceptions, you will almost always find that CoAP is sent over UDP and MQTT runs over TCP. The following sections will take an in-depth look at CoAP and MQTT. The CoAP Action Program (CoAP) grew out of the efforts of the IETF Constrained RESTful Environments (CoRE) working group to create a framework for resource-oriented constraint nodes and networks. The CoAP framework defines a simple and flexible way to operate sensors and actuators for data or control devices. The CoAP messaging standard is primarily designed to facilitate the exchange of messages between UDP endpoints, including Secure Transport Protocol, Data Transport Layer Security (DTLS). From a formatting perspective, CoAP messages consist of a short-length header field (4 bytes), a variable-length but required token field (0-8 bytes), and an options field (if required) and payload fields.

CoAP language format is simple and flexible. It allows CoAP to provide the low overhead required for network collaboration while being easy to verify and execute from limited resources.

CoAP can operate on IPv4 or IPv6. However, it is recommended to place the message in an IP packet and UDP data to avoid fragmentation. The default MTU size for IPv6 is 1280 bytes with no whitespace allocation allowed, and the maximum CoAP message size can reach 1152 bytes including the 1024-byte payload. For IPv4, applications need to limit themselves and set the IPv4 Fragmentation (DF) bit to be more efficient due to the possibility of IP fragmentation in the network. CoAP communication in IoT infrastructure can be implemented in various ways. The connection can be between devices on the same network, otherwise different, or between devices and the Internet or cloud server, all operating over IP. Proxy mechanisms have also been defined, and RFC 7252 specifies a simple HTTP protocol for CoAP. Since both HTTP and CoAP are IP-based protocols, the domain name can be anywhere on the network and does not need to be on the border between restricted and unrestricted connections.

Like HTTP, CoAP is based on REST architecture, but there is a "thing" that acts as a client and a server. By exchanging asynchronous messages, the client works through the server's rules. Identifies the Uniform Resource Identifier (URI) resource localised on the server. The server responds with a country message that will include the source designation. CoAP request/response semantics includes GET, POST, PUT, and DELETE methods.

Message Queue Telemetry Transport (MQTT) In the 1990s engineers at IBM and Arcom (acquired by Eurotech in 2006) were looking for a reliable, portable and cost-effective way to build server site monitors and manages Widely used in the oil and gas industry Many sensors and their data, such as those used. The result of their research is the development and implementation of the Message Queue Telemetry Transport (MQTT) protocol, which is now the standard established by the Organization for the Advancement of Information Systems (OASIS).

Client/server selection and broadcast/record functions based on TCP/IP architecture

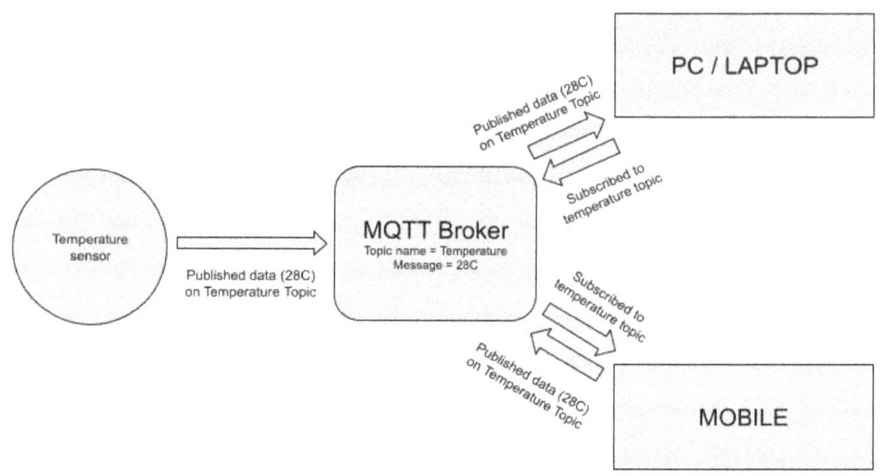

Exhibit 3.8 MQTT

An MQTT client can act as a publisher by sending data (or potential data) to an MQTT server that acts as an MQTT message broker. In the example, the MQTT client on the left is a temperature (Temp) and relative humidity (RH) sensor that reports Temp/RH data. The MQTT server (or message broker) accepts network connections and requests from broadcasters, such as temperature/relative humidity data. It also manages the registration and deregistration process and forwards data requests as a client to the MQTT client. The application on the right side of is an MQTT client, which is a client for temperature/relative humidity data generated by the transmitter or sensor on the left. This trend is well known as consumers want to receive information from the advertiser. Collaboration and social networking on Twitter is a good example. Through MQTT, clients can save all data (using the original text) or specific data in the published data tree. Additionally, the presence of message brokers in MQTT separates the data transfer between the client as the publisher and the client. In fact, publishers and subscribers do not know (or do not want to know) each other. The benefit of this termination is that the MQTT message broker ensures that messages are intact and cached while the network is down. This also means advertisers and subscribers don't need to be online at the same time. MQTT handles packets over TCP transport using port 1883. TCP provides a regular, unstructured stream of bytes between the MQTT client and the MQTT server. Alternatively, MQTT can be secured using TLS on port 8883 or Web Socket (defined in RFC 6455) can be used. MQTT is a lightweight protocol because each control packet has a fixed 2-byte header with configurable transport option and payload option. You should note that control packets can carry up to 256 MB of data. Figure 2.23 provides an overview of the MQTT message format.

Compared to the CoAP message format, MQTT has a smaller size of 2 bytes compared to the 4 bytes size of CoAP. The first MQTT field in the header is the message type, which identifies the type of MQTT packet in the message. There are 14 different types of control packets defined in MQTT version 3.1.1. Each one has a unique value hardcoded in the Message Type field. Note that the values 0 and 15 are reserved.

Keynotes:

Topic	Keynotes
IoT Fundamentals	The Internet of Things (IoT) revolutionizes connectivity, offering vast potential in domains like smart cities, healthcare, agriculture, and industry.
Network Technologies	Understanding OSI model layers and protocols is crucial for developing robust IoT networks.
IoT Networking	LPWAN, Bluetooth, ZigBee, NFC, RFID, WiFi, and Ethernet are among the key technologies shaping IoT communication.
Challenges	IoT networking faces challenges like range, bandwidth, power efficiency, intermittent connectivity, and interoperability.
Reference Model	The IoT reference model aids in understanding the layered architecture and promotes interoperability.
Ecosystem Overview	The IoT ecosystem comprises interconnected devices, networks, platforms, and agents, facilitating data collection and utilization.
IEEE 802.15.4 and ZigBee	Industrial wireless sensor networks leverage standards like IEEE 802.15.4 and protocols like ZigBee for efficient communication.

Trivia Challenge Industry 4.0:

Welcome to the Industry 4.0 Trivia Challenge! Get ready to test your knowledge of the wacky world of Industry 4.0 with these hilarious multiple-choice questions. Each question is paired with some side-splitting facts and figures about Industry 4.0. Let us dive in and have a giggle!

1) What's the secret sauce behind IIoT communication networks?

> A) The power of positive thinking...just kidding, it's all about those Wi-Fi signals!
>
> B) Advanced encryption algorithms to protect data from cyber hackers.
>
> C) Carrier routers?
>
> D) Avoiding technology altogether?

2) What is the ultimate goal of IIoT communication networks?

> A) To make data exchange as slow as molasses... said no one ever!
>
> B) To optimize communication for seamless connectivity and data exchange.
>
> C) To ignore technological advancements because who needs efficiency anyway?
>
> D) To exclude smartphones because they're just too smart for their own good!

3) What's the funniest thing about IIoT communication networks?

> A) They're powered by unicorn magic...just kidding, it's all about those data packets!
>
> B) They optimize communication for efficient data transfer, making the internet a happier place... almost.
>
> C) They focus on data carrier.
>
> D) They exclude modern technology altogether?

4) What's the ultimate goal of IIoT communication networks beyond large enterprises?

> A) To limit accessibility to advanced technologies.
>
> B) To democratize IIoT technologies.
>
> C) To exclude small and medium-sized enterprises because they're just not tech-savvy enough.
>
> D) To prioritize carrier pigeons because who needs Wi-Fi when you've got birds?

5) How do IIoT communication networks handle data privacy concerns?

A) Relying on carrier pigeons: Who needs data privacy when you've got birds?

B) By implementing robust encryption and access control measures...almost like a digital fortress!

C) By avoiding data altogether because who needs privacy anyway?

D) By excluding data privacy altogether because who needs secrets anyway?

6) What's the most surprising thing about IIoT communication networks?

A) They're powered by hamsters running on wheels...just kidding; it's all about those data centres!

B) They optimize communication for seamless connectivity and data exchange, making the internet a happier place...almost.

C) They focus on carrier pigeons because who needs routers when you've got birds?

D) They exclude modern technology altogether because who needs progress anyway?

7) How do IIoT communication networks ensure interoperability?

A) By hoping for the best and crossing their fingers. I'm just kidding. It's all about standardization and open protocols!

B) By avoiding interoperability altogether because who needs compatibility anyway?

C) Relying on carrier pigeons: Who needs interoperability when you've got birds?

D) By excluding interoperability measures altogether because who needs collaboration anyway?

8) What's the biggest challenge in implementing IIoT communication networks?

A) Convincing carrier pigeons to work overtime because who needs sleep when you've got birds?

B) Overcoming technological complexity and ensuring seamless integration...almost like untangling a giant ball of yarn!

C) Avoiding technological advancements altogether because who needs progress anyway?

D) Excluding implementation challenges altogether because who needs challenges anyway?

9) How do IIoT communication networks handle network congestion?

A) By hoping for the best and crossing their fingers—just kidding. It's all about traffic management and prioritization!

B) By avoiding network congestion altogether because who needs efficiency anyway?

C) Relying on carrier pigeons: Who needs network congestion when you've got birds?

D) By excluding congestion management altogether because who needs smooth data flow anyway?

10) What's the biggest obstacle to achieving seamless connectivity in IIoT communication networks?

A) Convincing carrier pigeons to work together because who needs teamwork when you've got birds?

B) Overcoming geographical barriers and ensuring consistent connectivity... almost like herding cats!

C) Avoiding connectivity altogether because who needs communication anyway?

D) Excluding seamless connectivity altogether because who needs reliability anyway?

Summary:

This Chapter delves into the Basics of Communication Networks in the context of the Industrial Internet of Things (IIoT). It begins by elucidating the fundamentals of IIoT and its wide-ranging applications. Subsequently, it explores network fundamentals, OSI model layers, and various IoT networking technologies such as LPWAN, Bluetooth, ZigBee, NFC, and WiFi. The chapter highlights considerations and challenges in IoT networking, emphasizing factors like range, bandwidth, power efficiency, and interoperability. Additionally, it discusses the IoT reference model and elucidates the IoT ecosystem's essential components. Lastly, it examines specific technologies like IEEE 802.15.4 and ZigBee in the context of industrial wireless sensor networks.

Abbreviation

Abbreviation	Term
IIoT	Industrial Internet of Things (IIoT)
IoT	Internet of Things (IoT)
OSI	Open Systems Interconnection (OSI)
LPWAN	Low Power Wide Area Network (LPWAN)
BLE	Bluetooth Low Energy (BLE)
NFC	Near Field Communication (NFC)
RFID	Radio Frequency Identification (RFID)
LAN	Local Area Network (LAN)
WAN	Wide Area Network (WAN)
PAN	Personal Area Network (PAN)
MAN	Metropolitan Area Network (MAN)
IEEE	Institute of Electrical and Electronics Engineers (IEEE)

Chapter 4

Augmented /Virtual Reality: Industry 4.0

Learning Objectives:

- Understanding Augmented Reality (AR) and Virtual Reality (VR)

- Define AR and VR technologies and distinguish between them.

- Explain how AR enhances real-world environments with digital information, while VR immerses users in entirely digital environments.

- Applications of AR and VR in Various Industries

- Explore the diverse applications of AR and VR in industries such as gaming, entertainment, healthcare, education, and training.

- Analyse the impact of AR and VR on user experience, interactivity, and extended reality (XR) integration.

- Development Process of AR Apps

- Identify the steps involved in developing AR applications, including planning, design, development, content creation, deployment, and maintenance.

- Understand the benefits of immersive experiences in enhancing operational efficiency, driving innovation, and unlocking new value propositions in Industry 4.0.

- Exploring the Potential of AR/VR Beyond Boundaries

4.0 Introduction AR & VR

The rise of Virtual Reality (VR) and Augmented Reality (AR) signals a complete turnaround in the ever-evolving Industry 4.0 world, where technology is king. These immersion technologies offer a transformative experience by seamlessly blending the boundaries between the physical and digital realms, revolutionising how individuals interact with their environment. In Industry 4.0, the appeal of AR and VR lies in their ability to enhance productivity, efficiency, and safety by providing immersive training simulations, real-time data visualisation, and remote assistance capabilities. However, in the context of Industry 4.0, what precisely is the appeal of AR and VR, and how are they changing the nature of work in the future?

Immersion technologies, like AR and VR, are more than simply catchphrases; they are forces for transformation that are redefining workflows and conventional industrial processes. With AR, workers may view and engage with their surroundings via a new lens by superimposing digital information over the real world. Imagine technicians who wear smart glasses with augmented reality capabilities, easily obtaining instructions in real time and seeing intricate machinery right before their eyes.

With augmented reality (AR), the tedious tasks of maintenance and troubleshooting Augmented Reality and Virtual Reality (AR/VR)

Augmented Reality (AR)

Definition: Augmented Reality places digital information into a real-world environment to enhance the use of one's perception of images. surrounds the environment.AR combines physical and digital spaces to provide real-time information content.

Features: AR enhances the user's existing environment by adding digital content. It does not change reality but enhances it, facilitating the seamless integration of virtual and physical content.

Virtual Reality (VR)

Definition: Virtual reality immerses the user in a completely digital environment, creating a simulated reality that separates them from the physical world therein. VR often involves the use of headsets to transport the user into a computer-generated environment.

Features: VR provides an immersive experience by replacing the real world with a computer-generated world. Users can interact with this production environment as if it were real, engaging all their senses for a better experience.

4.0.1 Applications in various industries

Games and entertainment

AR: AR in games supports the real world through digital media. Games like Pokémon Go use AR to provide fictional characters about the player's environment.

VR: VR is changing gaming by providing immersive experiences. Users can explore the virtual world, interact with objects, and experience games in an unprecedented way.

Healthcare

AR: Surgeons use AR to transfer patient information or medical images to the surgical site during surgery, improving accuracy. AR is also used in medical simulation.

VR: VR is used for medical treatment, therapy and training of doctors in a virtual environment.

Education and Training

AR: Educational apps use AR to provide interactive experiences. For example, AR can bring books to life by overlaying 3D models or additional information.

VR: VR enables training events such as virtual flight tests for pilots or real-life training in various industries.

Impact on user experience

AR: Users experience the integration of the digital and physical worlds, enhancing their understanding of the real world through additional information.

VR: Users enter a computer-generated environment, providing a sense of presence and interaction in a virtual environment.

Enhanced Interactivity

AR: Users can interact with digital content embedded in the real world, encouraging collaboration and discussion.

VR: Interactive VR is wide-ranging and allows users to control virtual objects, move around and experience a better understanding of existence. Next Trends and Innovations AR/VR in Social Engagement Integrating AR/VR into social media and communication platforms for greater engagement.

Provide virtual space for meetings and events, enhance remote collaboration.

4.0.2 Extended Reality (XR) Integration

The development of XR combines AR, VR and other immersive technologies to provide a seamless experience.

Integrating XR into industries such as construction allows professionals to visualise and understand interactions with 3D models.

As we delve deeper into the future of AR/VR, it is clear that this technology is not limited to certain sectors but has the potential to change the way we see and interact with the world. As innovation continues to expand, the distinction between reality and virtual reality is blurring, paving the way for a new era of experience and transformation, becoming dynamic, intuitive experiences that increase productivity and accuracy to levels that were previously difficult to achieve.

Aspect	Augmented Reality (AR)	Virtual Reality (VR)
Immersion	Integrates digital content into the real-world view.	Creates a fully immersive digital environment.
Interaction with Reality	Enhances real-world experiences with digital overlays.	Simulates entirely virtual environments, detached from reality.
Device Requirements	Typically accessed through smartphones or AR glasses.	Requires specialised VR headsets and controllers.
Applications	Commonly used in gaming, navigation, and education.	Widely employed in gaming, simulations, and training.
Spatial Awareness	Relies on understanding real-world spatial context.	Focuses on creating and manipulating virtual environments.

Exhibit 4.1 AR Vs VR

In the meantime, virtual reality (VR) immerses users in fully virtual settings where they can explore, manipulate, and engage with computer models and simulations.

Within Industry 4.0, virtual reality (VR) emerges as a potent instrument for remote collaboration, immersive training, and design prototyping. Imagine engineers working together from different countries, engaging with 3D prototypes with ease as though they were in person. Virtual reality (VR) blurs the lines between physical space and enters in a new era of international cooperation and creativity.

However, the benefits of AR and VR go far beyond increased productivity, they open up a world of limitless opportunities in a variety of industries. AR and VR technologies are transforming patient care and medical education in the healthcare industry by providing tools for visualisation and immersive simulations that raise the bar for quality. AR-enabled shopping experiences are revolutionising the way consumers engage with products in retail, effectively bridging the gap between the digital and real worlds. AR and VR are changing industries, opening up new opportunities for development and discovery in everything from education to entertainment.

However, there are obstacles in the way of fully utilising AR/VR's potential in Industry 4.0. Barriers related to organisations and cultures must be addressed in addition to technical issues like scalability and interoperability. Businesses need to get leadership support for AR/VR projects, coordinate them with strategic goals, and promote a collaborative and experimental culture. Then only they will be able to fully realise AR/VR's transformative potential and advance into the workforce of the future.

The use of AR and VR in Industry 4.0 will continue to change the fundamentals of manufacturing and industrial processes as we set out on this path of invention and discovery. It's a voyage full of limitless opportunities, where the boundaries between truth and fantasy are blurred and the future appears before our own eyes.

Businesses who embrace AR and VR as essential elements of their digital transformation journey are paving the path for a more productive, sustainable, and interconnected society where innovation has no boundaries and the only restrictions on what may be imagined.

4.1 Process of Developing Augmented Reality Apps

The creation of successful augmented reality experiences involves a set of steps in the augmented reality development process.

Planning: Creating a vision, researching rivals, determining target users, and outlining hardware specifications are all part of the first step. What interactions and experiences will be most beneficial to users is determined by researchers.

Design is the next step, where developers set up user interfaces, select AR sensors and backends, and define the functionality of augmented reality apps. Designers can test concepts through prototyping before moving on with development.

Development: Included in the AR development stage are:

- Programming the application
- Making animations and 3D models
- Combining Sensor APIs and AR
- Linking up with necessary backends

- Testing guarantees that parts function as intended.

Content Creation: To improve the augmented reality experience, content is produced after development. Text, images, audio, movies, animations, and 3D models are all included in this. Experiences are brought to life using AR content.

Deployment: As an enterprise solution or in the appropriate app stores, the AR app is made available. You can add more stuff as needed.

Maintenance: The AR software has to have features added, issues fixed, and performance improved after it is released. Further development of AR is guided by user feedback.

4.2 Process of Developing Virtual Reality Apps

A multi-step process including strategy, design, development, content creation, deployment, and maintenance is required to create a successful VR app.

Planning: In the first phase, developers define their goals, research the competitors, and determine who their target audience is. Which VR experiences will be interesting, useful, and innovative are determined by researchers. There are specific hardware requirements.

Design: Design consists of

- Describing the augmented reality app's features

- Setting up virtual reality user interfaces

- Choosing Virtual Reality Sensors

- Selecting backend remedies

Before developing a virtual reality app, designs may be tested early on thanks to wire framing and prototyping.

Development: Development includes writing the app's code, making 3D animations and models, incorporating VR SDKs and APIs, and establishing connections to the necessary backends. Testing guarantees that parts function as intended in the virtual environment.

Content Creation: To finish the virtual environment, 3D models, animations, music, films, images, and text are produced. Content that is captivating and realistic helps make the VR experience come to life.

Deployment: App stores and appropriate VR platforms get the deployment of the app. Distribution for enterprise solutions may be more tailored.

Maintenance: To add features, correct bugs, enhance performance, and accommodate new VR devices, post-release VR apps need to be maintained. Ongoing development of virtual reality apps is guided by user input.

Usability testing: Representative users assist in identifying problems and ensuring that the VR experience is user-friendly, engaging, and beneficial at every significant stage. Users give prototypes a try and offer comments.

Accessibility: The VR app's creators take into account how to accommodate users with physical, cognitive, auditory, and visual impairments from the outset. Better augmented reality applications are made possible by solutions that are incorporated into the design.

Publication and Marketing: After the VR app is released, its visibility and user acquisition depend heavily on marketing it with targeted advertisements, tech media coverage, social media posts, and VR app store promotion. Essential Tools and Technologies for Developing AR and VR Apps

For AR/VR development to provide useful and captivating experiences, a variety of VR prototyping tools and technologies are needed. Here are some of these points:

Hardware

VR headsets such as the Oculus Rift, HTC Vive, and Valve Index offer immersive VR experiences with wide field-of-view screens, high-resolution panels, and motion-tracking sensors. AR headsets, such as Microsoft HoloLens, Meta 2, and Magic Leap, use see-through lenses or displays, spatial mapping sensors, and gesture tracking to overlay digital items in the real environment. Motion sensors such as accelerometers, gyroscopes, and magnetometers monitor headgear movement and orientation, allowing interactivity.

Sensor Technologies

Computer vision technology such as RGB cameras, depth cameras, and inertial measurement units (IMUs) enable features such as object recognition, spatial mapping, and hand/finger tracking to create realistic AR and VR experiences. Simultaneous localization and mapping (SLAM) techniques use sensor data to map virtual objects onto natural settings in real time, allowing them to remain in physical space.

SDKs

SKDs provide APIs for key AR and VR capabilities. SDKs such as ARCore, ARKit, Unity, Unreal Engine, and Vuforia provide graphics rendering, spatial mapping, object identification, and other capabilities. SteamVR and OpenVR are software development kits for virtual reality games and applications Languages

Languages

AR/VR apps are developed using programming languages such as C++, C#, JavaScript, Python, and Unity's C#. C++ and lower-level languages offer performance and control for graphics rendering and time-critical activities, whilst higher-level languages allow for faster prototyping.

3D Modelling

To fill virtual worlds, designers can create high-fidelity 3D models of virtual items, characters, and locations with the aid of 3D modelling software like Blender, 3DS Max, and Maya.

4.3 Best Practices for Augmented Reality App Development

To achieve great performance, usability, comfort, and an interesting user experience, VR and AR software developers must adhere to a set of best practices. Here are the most crucial practices to follow when developing these apps.

Prioritise the user experience. Ensure that programmes are user-friendly and beneficial. Provide users with memorable and entertaining experiences.

Maintain high performance levels. For pleasant VR experiences and reduced latency, a minimum of 60 frames per second is recommended. Optimise the code and assets. Use asset compression. Compress 3D models, augmented reality frameworks, and sounds to minimise app size while maintaining quality.

Use adaptive rendering. To sustain frame rates, adjust graphics settings dynamically based on performance.

Test all target hardware. To spot bugs early on, test apps on all VR and AR headsets and devices that will use them.

Perform extensive testing. Before releasing any features, test them in numerous usage situations to find problems and vulnerabilities. Iterate through testing.

Include load displays. To avoid sickening users with stutters, display a loading screen while large assets are loaded.

Support multiple modalities of interaction. Consider gaze, gesture, haptic, and controller inputs for greater flexibility.

Reduce abstraction. Remove any layers between users and virtual objects to improve responsiveness. Support 6DOF. Allow 6 degrees of freedom (directional movement and rotation) for realistic interactivity.

Provide haptic feedback. Use vibrations, sound, and other effects to correspond to users' actions in the virtual environment.

Design for comfort. Limit VR experiences to 10-15 minutes until consumers establish their VR "legs." Avoid "janky" motions.

Provide a VR comfort rating. Inform users about potential motion sickness, and tailor content to their comfort levels.

Prepare for hardware upgrades. Create future-proof technologies to support the expanding

4.3.1 AR and VR devices

Consider accessibility. Allow users with physical, mental, hearing, and vision disabilities to utilise the applications.

Collect user input. Actively seek feedback and suggestions to enhance the user experience and find faults. Launch and then iterate. Launch apps "out of the door" and keep refining them in response to user input and actual use.

When creating an AR or VR application, major priorities should include ensuring a seamless and responsive user experience, high performance, device compatibility, user comfort, and accessibility. Over time, apps can be made better by gathering feedback and iterating continuously after launch.

Challenges and Limitations of AR and VR App Development: - Hardware limitations

Current VR and AR headsets have drawbacks such as tiny fields of view, low resolution, limited range of motion, and exorbitant prices. This prevents developers from completely realising their ambitions. Hardware is fast improving, but it still needs to reach a level that allows for completely realistic and immersive experiences.

Processing Power

Complex 3D visuals must be rendered in real time, which requires powerful CPUs and graphics cards. However, many mobile devices lack the processing power and augmented reality foundation required for high-quality AR, necessitating application performance optimisation. Input latency

Even minor delays between users' bodily actions and what they perceive in VR/AR can create pain and undermine the illusion of realism. Reducing input latency is a technical challenge.

Battery Life

Particularly for mobile devices, the sensors, fast refresh rates, and graphics rendering needed for AR/VR greatly deplete batteries, which limits the design of applications.

Ergonomics

Heat, pressure points, and weight distribution are all elements that can make current headsets uncomfortable to use for long periods of time. Ergonomic considerations restrict ideal usage times.

Technological Silos

Cross-platform development and interoperability are hampered by the closed ecosystems in which many VR/AR platforms function, complete with proprietary SDKs and specs.

Lack of Standards

The AR/VR market requires uniform standards for graphics, interfaces, objects, and interactions, which makes it difficult for experiences to work across multiple platforms.

Software Bugs and Glitches

Software flaws in complex AR/VR apps can lead to errors and undermine the perception of a genuine, integrated experience. Thorough testing helps to reduce problems.

Network Performance

When network performance is stable, relying on it for cloud services and social features can harm the AR/VR experience.

The current state of technology makes it difficult for developers to create compelling and completely immersive AR and VR experiences due to restrictions in hardware, battery life, input latency, ergonomics, lack of standards, and network speed. Though technology is continuously advancing, obstacles must be overcome that limit the potential of applications.

4.4 New Trends in AR and VR Application Development

AR and VR technology are fast evolving, creating new opportunities for immersive experiences. Several developing themes are influencing the future of AR and VR app development.

Increased Hardware Performance - AR and VR headsets are getting more advanced, with greater resolutions, larger fields of view, slimmer designs, and better tracking. This allows for more lucrative, realistic virtual experiences with less latency and jitter.

Enhanced Mobile Chipsets: Advanced mobile processors and graphics cards provide the computational capacity required for enhanced augmented reality experiences on smartphones and tablets. This could make augmented reality more accessible to general users.

5G Networking - The deployment of 5G networks will provide the bandwidth and low latency needed for cloud-based AR/VR and high-fidelity 3D graphic streaming. This could open up new social and multiplayer applications. Improved sensors for eye and hand tracking will enable more intuitive interactions in AR and VR. Applications can respond to a user's gaze direction and hand motions, providing a more natural experience.

Unified AR and VR Platforms - Many firms are developing platforms that combine AR and VR experiences into a single software stack. This will make it easier for developers to design applications that combine elements of augmented and virtual reality.

Mixed Reality Continuum - As technology advances, the distinction between AR and VR becomes less clear. Applications will combine parts of both to create varying degrees of immersion based on user preferences and context. Improved Haptics - Advancements in haptic technology seek to reproduce the sense of touch in the virtual world. Advances such as haptic gloves and suits have the potential to make VR and AR interactions far more lifelike.

More augmented reality applications are emerging, including multiuser AR and VR applications that allow users to share immersive experiences from various locations. This might revolutionise communication, education, and collaboration.

Over the next ten years, these trends all suggest that AR and VR technology will become more smoothly incorporated into our daily lives and jobs. As hardware limits are overcome and platforms converge, we may expect an explosion of inventive applications that alter how we use technology.

4.5 Enhancing Efficiency: AR/VR's Impact on Manufacturing

AR/VR technologies are revolutionising the manufacturing industry by profoundly enhancing efficiency across all stages of the manufacturing lifecycle. These immersive technologies are not just augmenting reality; they are transforming the way manufacturers conceive, design, produce, and maintain their products.

Beginning with the product design phase, AR/VR solutions facilitate a collaborative and iterative approach, allowing designers and engineers to visualise and interact with 3D models in a virtual environment. This capability expedites the prototyping process, enabling faster iteration cycles and more accurate feedback, ultimately leading to reduced time-to-market for new products. Furthermore, AR/VR simulations enable manufacturers to conduct virtual testing and validation, identifying potential design flaws or assembly issues before physical prototypes are built, thus minimising costly revisions and rework.

Moving onto production planning and execution, AR/VR technologies offer invaluable tools for optimising manufacturing processes and workflows. By overlaying digital information onto physical spaces, AR enables workers to receive real-time instructions and guidance directly within their field of view, streamlining assembly tasks and reducing human error. Similarly, VR-based training simulations provide a risk-free environment for workers to familiarise themselves with complex machinery and procedures, accelerating the on boarding process and ensuring consistent quality standards across the workforce. Moreover, AR/VR-driven production monitoring and analytics allow supervisors to track key performance metrics in real-time, identify bottlenecks, and implement timely adjustments to improve overall efficiency and throughput.

Quality control is another area where AR/VR technologies are making a significant impact. Traditional quality inspection processes often rely on manual methods that are time-consuming and prone to errors. However, AR-enabled smart glasses equipped with computer vision capabilities can overlay digital overlays onto physical objects, highlighting defects or deviations from specifications in real-time. This enhances the accuracy and efficiency of quality inspections, ensuring that only products meeting stringent quality standards are shipped to customers. Additionally, VR-based simulations enable inspectors to immerse themselves in virtual replicas of manufacturing environments, facilitating more comprehensive and detailed inspections than would be possible in the physical world.

4.6 Beyond Boundaries: Exploring the Vast Potential of AR/VR

The integration of Augmented Reality (AR) and Virtual Reality (VR) technologies within the framework of Industry 4.0 represents a seismic shift that extends far beyond the traditional confines of manufacturing. Across diverse sectors including healthcare, retail, education, and entertainment, the transformative influence of AR/VR is reshaping the landscape of business operations, consumer experiences, and educational methodologies.

In healthcare, AR/VR technologies are catalysing significant advancements in medical training, patient care, and surgical procedures by providing immersive simulations and visualisation tools that enable medical professionals to hone their skills, refine their techniques, and enhance their decision-making processes.

Through lifelike simulations and realistic scenarios, practitioners can effectively train for complex procedures, improve diagnostic accuracy, and ultimately elevate the standard of patient care. Similarly, in the realm of retail, AR-powered shopping experiences and virtual try-on solutions are revolutionising the way consumers interact with products, allowing them to visualise items in real-world settings, virtually try on clothing and accessories, and make informed purchasing decisions with confidence. By seamlessly integrating AR/VR technologies into the retail landscape, businesses can enhance customer engagement, drive sales, and differentiate themselves in an increasingly competitive market environment, ultimately redefining the retail experience in the digital age.

Aspect	Augmented Reality (AR)
Input Form	Typically utilises gestures, voice commands, touch, and motion tracking. Users interact with digital content overlaid onto the real world.
Output Form	Presents augmented content through various mediums such as smartphones, AR glasses, heads-up displays (HUDs), or projection systems. The augmented content is seamlessly integrated into the user's real-world view.
User-Device Interaction	Users engage with AR content by manipulating their devices, often through touchscreen gestures, voice commands, or physical movements. Interaction can also involve gaze-based selection and manipulation using AR glasses or headsets. AR devices track the user's movements and surroundings to deliver an immersive experience.

Exhibit 4.2 AR Elements

Furthermore, in the field of education, AR/VR-based learning platforms are disrupting traditional teaching methodologies by offering interactive and immersive educational experiences that cater to diverse learning styles and preferences. Through virtual field trips, interactive simulations, and hands-on activities, students are empowered to explore complex concepts, deepen their understanding, and engage with course material in meaningful ways. By leveraging AR/VR technologies, educators can foster student engagement, improve retention rates, and cultivate critical thinking skills, ultimately preparing students for success in an increasingly digital and interconnected world. Across all sectors, the full spectrum of AR/VR applications unlocks new opportunities for industries to innovate, evolve, and thrive in the Industry 4.0 era. By embracing these immersive technologies, businesses can tap into new revenue streams, develop innovative business models, and deliver enhanced value propositions to customers, ultimately driving growth, competitiveness, and sustainability in the Fourth Industrial Revolution and beyond.

4.7 Creating Immersive Experiences: The Integration of AR/VR in Industry 4.0 Practices

In the dynamic landscape of Industry 4.0, the successful integration of Augmented Reality (AR) and Virtual Reality (VR) demands a multifaceted approach that addresses technical, organisational, and cultural challenges. From a technical perspective, ensuring interoperability, scalability, and security of AR/VR systems is paramount. Companies must invest in robust infrastructure and technologies capable of supporting immersive experiences while safeguarding sensitive data. Additionally, adherence to industry standards and protocols is essential to facilitate seamless integration with existing systems and processes.

Organizationally, aligning AR/VR initiatives with strategic business objectives is imperative. Securing executive buy-in and support is crucial to allocate resources, prioritise initiatives, and drive adoption across the organisation. Establishing cross-functional teams comprising experts from various departments, including IT, operations, and marketing, fosters collaboration and ensures alignment between AR/VR initiatives and broader business goals. Moreover, creating a culture that embraces experimentation, innovation, and collaboration is essential for AR/VR success. Encouraging employees to explore new technologies, share ideas, and iterate on solutions cultivated a dynamic environment conducive to innovation and continuous improvement. By embracing a comprehensive approach to AR/VR implementation and adoption, companies can harness the power of immersive experiences to enhance operational efficiency, drive innovation, and unlock new value propositions in the rapidly evolving landscape of Industry 4.0.

4.8 Innovating Industry

As Industry 4.0 continues its evolution, Augmented Reality (AR) and Virtual Reality (VR) technologies are poised to become increasingly integral, shaping the landscape of innovation, differentiation, and competitive advantage. Their potential applications are vast and diverse, promising to enhance worker productivity and safety while paving the way for new business models and revenue streams. Within the context of Industry 4.0, the possibilities presented by AR/VR seem limitless. By fostering a culture of experimentation, agility, and collaboration, organisations can harness the transformative power of AR/VR to unlock untapped opportunities, surmount challenges, and flourish in the digital era. As we embark on this journey of innovation and exploration, the role of AR/VR in Industry 4.0 will continue to redefine the future of manufacturing and industrial processes, ushering in an era characterised by heightened efficiency, sustainability, and interconnectedness.

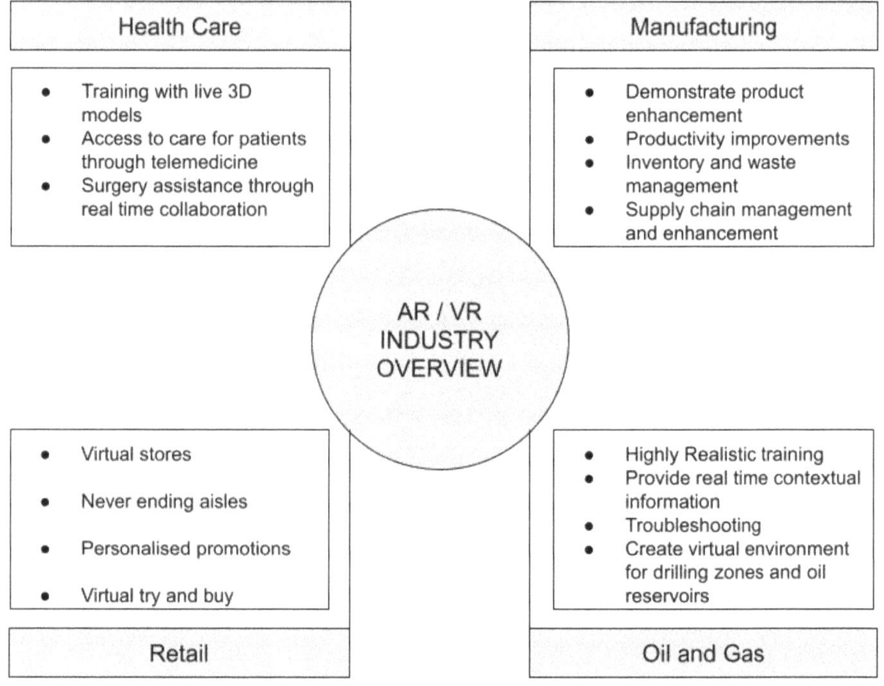

Exhibit 4.3 AR/VR Applications Industry Wise

The future of Industry 4.0 holds boundless opportunities for AR/VR technologies to continue reshaping manufacturing processes, paving the way for a more connected and efficient world.

Keynotes:

Topic	Keynotes
Augmented Reality (AR)	Augmented Reality places digital information into a real-world environment to enhance the use of one's perception of images. AR combines physical and digital spaces.
Virtual Reality (VR)	Virtual reality immerses the user in a completely digital environment, creating a simulated reality that separates them from the physical world.
Applications in various industries	- Games and entertainment - Healthcare - Education and Training - Impact on user experience - Enhanced Interactivity - Extended Reality (XR) Integration
Process of Developing AR Apps	- Planning - Design - Development - Content Creation - Deployment - Maintenance
Hardware, Sensor Technologies, SDKs, Languages, 3D Modelling	-
Best Practices for AR App Development	- AR and VR devices
New Trends in AR and VR Development	- Increased Hardware Performance - Enhanced Mobile Chipsets - 5G Networking - Unified AR and VR Platforms - Mixed Reality Continuum - Improved Haptics
Enhancing Efficiency: AR/VR's Impact on Manufacturing	AR/VR technologies are revolutionising the manufacturing industry by profoundly enhancing efficiency across all stages of the manufacturing lifecycle.
Beyond Boundaries: Exploring the Vast Potential of AR/VR	The integration of Augmented Reality (AR) and Virtual Reality (VR) technologies within the framework of Industry 4.0 represents a seismic shift.
Creating Immersive Experiences: The Integration of AR/VR in Industry 4.0 Practices	In the dynamic landscape of Industry 4.0, the successful integration of Augmented Reality (AR) and Virtual Reality (VR) demands a multifaceted approach.
Innovating Industry	As Industry 4.0 continues its evolution, Augmented Reality (AR) and Virtual Reality (VR) technologies are poised to become increasingly integral.

Trivia Challenge Industry 4.0

Welcome to the Industry 4.0 Trivia Challenge! Get ready to test your knowledge of the wacky world of Industry 4.0 with these hilarious multiple-choice questions. Each question is paired with some side-splitting facts and figures about Industry 4.0. Let us dive in and have a giggle!

1) What do AR and VR stand for in the context of Industry 4.0?

A) Amazing Realities and Vibrant Realms

B) Awesome Rockets and Volcanic Rumbles

C) Augmented Reality and Virtual Reality

D) Alluring Rainbows and Vivid Rainforests

2) How are AR and VR technologies described in the chapter?

A) As the magic wands of Industry 4.0, conjuring digital wonders.

B) As the secret sauce is sprinkled over factories, they taste like the future.

C) As the friendly ghost's haunt manufacturing halls, spooking inefficiency away.

D) As the time machines transporting workers back and forth between reality and cyberspace.

3) What is a key application area for AR and VR technologies discussed in the chapter?

A) Sending virtual postcards from the assembly line to customers.

B) Enhancing efficiency in manufacturing processes.

C) Hosting virtual reality tea parties for factory workers.

D) Creating augmented reality treasure hunts in warehouses.

4) What processes are involved in developing AR and VR apps according to the chapter?

A) Whispering magical incantations and hoping for the best.

B) Summoning tech-savvy unicorns to do all the work.

C) Delving into the mysterious realms of coding and application development.

D) Holding a séance to communicate with the spirits of tech pioneers.

5) **How does integrating AR and VR drive innovation across industries?**

   A) By transforming factories into virtual reality amusement parks.

   B) By hosting AR and VR costume parties for corporate events.

   C) By creating immersive experiences that push the boundaries of traditional practices.

   D) By teleporting workers to virtual beach resorts during lunch breaks.

6) **Which of the following statements is NOT true about AR and VR?**

   A) AR overlays digital information onto the real world.

   B) VR creates entirely virtual environments.

   C) AR stands for Amazing Robots, while VR stands for Virtually Real.

   D) Both technologies have applications beyond entertainment.

7) **How are new trends in application development described in the chapter?**

   A) Like shooting stars streaking across the digital sky.

   B) Like mushrooms sprouting after a technological rainstorm.

   C) Like rare Pokémon popping up in the augmented reality wilderness.

   D) Like diamonds waiting to be discovered in the virtual rough.

8) **What metaphor is used to describe the potential of integrating AR and VR?**

   A) Like merging reality and imagination to create a digital wonderland.

   B) Like mixing peanut butter and jelly to make the ultimate sandwich.

   C) Like playing a virtual reality game of chess with industry giants.

   D) Like bringing a splash of colour to the grayscale world of manufacturing.

9) **How does the chapter describe the impact of AR and VR on traditional boundaries?**

   A) Like breaking through the fourth wall of manufacturing.

   B) Like dissolving the barriers between physical and digital realms.

   C) Like transforming factory floors into virtual reality dance studios.

   D) Like turning warehouses into magical kingdoms of augmented enchantment.

10) In what way do AR and VR enhance efficiency in manufacturing according to the chapter?

    A) By teleporting workers to virtual coffee breaks on demand.

    B) By providing real-time data overlays for streamlined processes.

    C) By replacing human workers with virtual reality avatars.

    D) By hosting virtual reality team-building exercises on the factory floor.

Summary:

This Chapter delves into the realm of Augmented Reality (AR) and Virtual Reality (VR) within the context of Industry 4.0, highlighting their transformative potential across various industries. The chapter begins by defining AR and VR technologies, elucidating their distinctive features and applications. It explores how AR enhances real-world environments with digital overlays, while VR immerses users in entirely virtual environments.

The discussion extends to the development process of AR applications, outlining the steps involved from planning and design to deployment and maintenance. Key considerations such as hardware, sensor technologies, SDKs, programming languages, and 3D modelling are explored in depth. Moreover, the chapter elucidates best practices and emerging trends in AR/VR development, emphasizing the importance of technological advancements and industry standards.

Furthermore, the chapter underscores the profound impact of AR and VR on manufacturing efficiency, detailing how these immersive technologies optimize various stages of the manufacturing lifecycle. It also examines the broader implications of AR/VR integration beyond manufacturing, exploring their applications in healthcare, retail, education, and entertainment. Overall, Chapter 4 illuminates the vast potential of AR and VR technologies to revolutionize industries, reshape consumer experiences, and drive innovation in the digital age.

Abbreviations:

Abbreviations	Terms
AR	Augmented Reality (AR)
VR	Virtual Reality (VR)
XR	Extended Reality (XR)

Abbreviations	Terms
SDKs	SDKs: Software Development Kits
3D	3D: Three-Dimensional
IoT	IoT: Internet of Things
LAN	LAN: Local Area Network
WAN	WAN: Wide Area Network
MAN	MAN: Metropolitan Area Network
IEEE	IEEE: Institute of Electrical and Electronics Engineers
5G	5G: 5th Generation Wireless Technology
Haptics	Haptics: Touch Feedback Technology
AR/VR	AR/VR: Augmented Reality/Virtual Reality

Chapter 5

Blockchain Industry 4.0

Learning Objectives:

- Understanding Blockchain Fundamentals

- Define the core principles of blockchain technology, including decentralization, immutability, and cryptographic security.

- Explain how blockchain functions as a distributed ledger and its role in ensuring trust and transparency in transactions.

- Exploring Blockchain Applications in Industry 4.0

- Identify key use cases of blockchain technology in Industry 4.0, such as supply chain management, data integrity, and cybersecurity.

- Analyse the potential impact of blockchain on traditional industrial practices and business models.

- Assessing the Evolution of Blockchain

5.1 Introduction Blockchain

As Industry 4.0 continues its evolution, propelled by the relentless march of technological advancements, blockchain emerges as a transformative and disruptive force poised to reshape operations, instil trust, and show new business models across various sectors. This comprehensive exploration endeavours to shed light on the multifaceted role of blockchain within the context of Industry 4.0, illuminating its potential to revolutionise processes ranging from supply chains and data management to collaboration and beyond.

Positioned at the vanguard of innovation, blockchain transcends its role as a mere decentralised ledger, embodying a paradigm shift in how industries conceptualise and execute transactions, ensure security, and govern operations. At its core, blockchain represents a fundamental departure from centralised systems, offering a distributed and immutable ledger that engenders trust and transparency in transactions while mitigating the risks of fraud and manipulation. Through its ability to execute smart contracts autonomously, blockchain enables decentralised innovation, facilitating agreements and transactions without the need for intermediaries, thus streamlining processes and reducing costs across various domains.

Furthermore, blockchain's application in supply chain management heralds a new era of transparency and accountability, as it allows for the immutable recording of product provenance, ensuring the integrity of goods and mitigating the risks of counterfeiting and fraud. By providing a tamper-proof record of transactions, blockchain enhances supply chain visibility, enabling stakeholders to track the journey of products from origin to destination in real-time, thereby improving efficiency and customer satisfaction while reducing operational costs and risks. Beyond supply chains, blockchain has the potential to revolutionise data management practices, offering a secure and decentralised solution for storing, sharing, and verifying information. By leveraging cryptographic techniques and distributed consensus mechanisms, blockchain ensures the integrity and authenticity of data, safeguarding against unauthorised access, manipulation, and data breaches. Moreover, blockchain facilitates seamless collaboration and information exchange among stakeholders, enabling real-time access to trusted data while preserving data privacy and sovereignty. As industries grapple with the challenges of digital transformation, blockchain emerges as a catalyst for change, promising to redefine traditional industrial practices and providing advantages like efficiency, transparency, and resilience. Its disruptive impact extends far beyond the realms of technology, transcending industries and reshaping the way businesses operate and interact in the digital age.

As we delve deeper into the intricacies of blockchain technology, its transformative potential becomes increasingly apparent, paving the way for a future where trust is intrinsic, collaboration is seamless, and industries thrive amidst unprecedented digital transformation. By embracing blockchain, enterprises can unlock new opportunities for innovation, differentiation, and sustainable growth, positioning themselves at the forefront of the Fourth Industrial Revolution and shaping the future of industry for generations to come.

5.2 Blockchain Unleashed: Transforming Industry 4.0

Blockchain technology, once primarily associated with cryptocurrencies, has undergone a remarkable evolution, emerging as a formidable force in reshaping the landscape of Industry 4.0. At its essence, blockchain serves as a decentralised and immutable ledger, facilitating transparent, secure, and tamper-proof transactions and data exchange. Through the utilisation of cryptographic techniques and consensus mechanisms, blockchain effectively bypasses the necessity for intermediaries, resulting in reduced transaction costs and heightened data integrity.

Within the framework of Industry 4.0, blockchain's transformative potential is being realised across an array of sectors, unveiling unprecedented opportunities for automation, trust-building, and collaboration. Its decentralised nature fundamentally alters traditional transactional paradigms, fostering an environment conducive to innovation and efficiency. By leveraging blockchain technology, businesses can streamline processes, bolster trust among stakeholders, and foster collaborative efforts that transcend organisational boundaries. As the adoption of blockchain proliferates across industries, it is poised to serve as a catalyst for driving forward the evolution of Industry 4.0, ushering in an era of heightened connectivity, transparency, and resilience.

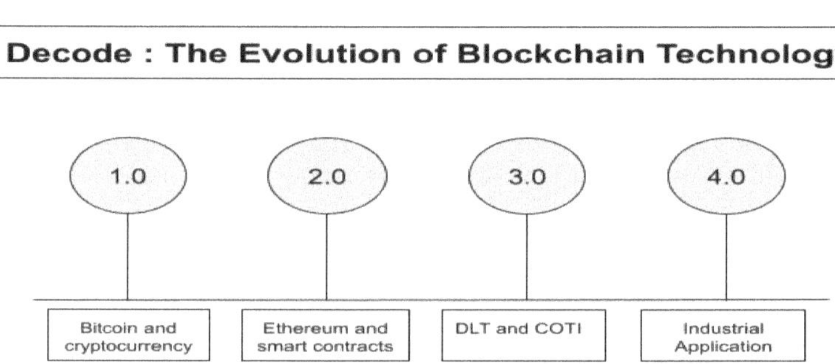

Exhibit 5.1 Evolution of Blockchain Technology

5.3 Decentralised Innovation: The Role of Blockchain in Industry 4.0

The decentralised nature of blockchain technology stands as a cornerstone feature, offering the potential for trustless transactions and peer-to-peer interactions that redefine the landscape of Industry 4.0.

Within this context, decentralised blockchain networks emerge as enablers of innovative solutions, presenting avenues for transformative advancements such as smart contracts, decentralised autonomous organisations (DAOs), and tokenized ecosystems.

Smart contracts, deployed on blockchain platforms, epitomise this innovation by embodying self-executing agreements that execute transparently and autonomously, eliminating the necessity for intermediaries. These contracts operate seamlessly, facilitating automated transactions and contractual obligations with unprecedented efficiency and reliability. In parallel, DAOs represent a paradigm shift in organisational structure, embodying decentralised governance models governed by code and community consensus. Through DAOs, collaborative decision-making and resource allocation are democratised, fostering inclusivity and transparency in organisational processes. By embracing decentralised innovation, Industry 4.0 stands poised to unlock new frontiers of efficiency, transparency, and inclusivity, heralding a future characterised by decentralised governance, automated agreements, and collaborative ecosystems that transcend traditional boundaries.

5.4 Building Trust: Blockchain's Impact on Industry 4.0

Trust stands as a foundational pillar of Industry 4.0, serving as the bedrock upon which seamless collaboration and value exchange among stakeholders are built. However, traditional trust mechanisms inherent in centralised systems often prove to be inefficient, prone to manipulation, and inadequate for the dynamic and interconnected nature of modern industries. Enter blockchain technology—a disruptive innovation poised to address these challenges head-on by providing a decentralised and immutable ledger that revolutionises the way trust is established and maintained in the digital realm. At its core, blockchain ensures trust through cryptographic verification and consensus mechanisms, rendering transactions and data immutable and transparent while eliminating the need for intermediaries or trusted third parties. This transformative capability finds its most poignant application in supply chain management, where blockchain enables end-to-end visibility and traceability of goods, from the sourcing of raw materials to the final delivery of finished products.

By recording each transaction and transfer of ownership on an immutable ledger, blockchain reduces the risk of counterfeiting, fraud, and supply chain disruptions, thereby enhancing trust and integrity across the entire supply chain ecosystem. Similarly, in the realm of data management, blockchain emerges as a beacon of trust by providing a secure and transparent framework for storing, sharing, and monetizing data. Through cryptographic techniques and distributed consensus mechanisms, blockchain ensures the integrity and authenticity of data while preserving privacy and ownership rights—a feat that was previously unattainable with centralised data storage systems. By leveraging blockchain, businesses can establish trust in the veracity and provenance of data, enabling seamless collaboration and information exchange among stakeholders while mitigating the risks of data breaches, unauthorised access, and manipulation.

Furthermore, blockchain's impact on trust extends beyond supply chains and data management, permeating various facets of Industry 4.0 and catalysing greater collaboration, innovation, and resilience in the face of evolving challenges. By building trust through blockchain, Industry 4.0 can unlock new opportunities for value creation, drive efficiencies, and foster a culture of transparency and accountability that underpins sustainable growth and competitiveness in the digital age.

As industries continue to embrace the transformative potential of blockchain technology, the establishment of trust emerges as a fundamental enabler of progress, propelling Industry 4.0 towards a future where collaboration knows no bounds, innovation flourishes, and trust is intrinsic to every transaction and interaction. In this paradigm shift towards trust-centric ecosystems, blockchain serves as a catalyst for positive change, redefining the way industries operate, interact, and thrive in the digital era.

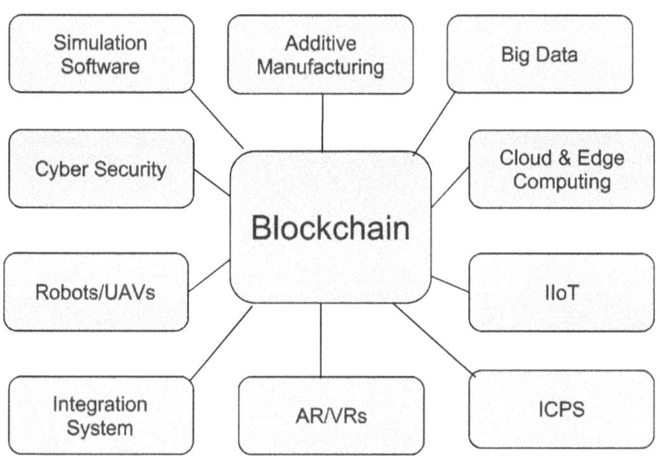

Exhibit 5.2 Blockchain Elements

5.5 Securing the Future: Blockchain Technology in Industry 4.0

As the evolution of Industry 4.0 unfolds, cybersecurity looms large as a critical concern, with the threat landscape becoming increasingly complex and pervasive. Conventional cybersecurity measures such as firewalls, antivirus software, and encryption, once stalwarts in the defence against digital threats, are now deemed insufficient to combat the emergence of sophisticated adversaries wielding tactics like ransomware, data breaches, and insider attacks. In this shifting landscape, blockchain technology emerges as a beacon of hope, offering a paradigm shift in cybersecurity by furnishing a decentralised and tamper-proof framework for safeguarding digital assets and transactions. Through its array of features including cryptographic hashing, consensus mechanisms, and distributed consensus, blockchain stands as a bulwark enhancing the resilience and integrity of digital ecosystems within the realm of Industry 4.0.

By decentralising data storage and transaction validation, blockchain mitigates the vulnerabilities inherent in centralised systems, thereby reducing the risk of single points of failure and unauthorised access. The cryptographic hashing employed by blockchain ensures that data remains immutable and tamper-proof, safeguarding against malicious alterations and unauthorised modifications. Moreover, consensus mechanisms such as proof of work or proof of stake enable network participants to collectively validate transactions and maintain a shared ledger of immutable records, further fortifying the integrity and transparency of digital transactions.

Additionally, the distributed nature of blockchain networks ensures that no single entity can exert undue influence or control over the system, fostering a trustless environment where transactions are executed with confidence and security. In the context of Industry 4.0, where digital assets and transactions abound across interconnected networks of devices and systems, the adoption of blockchain technology holds profound implications for cybersecurity. By leveraging blockchain's decentralised architecture and cryptographic safeguards, Industry 4.0 can mitigate cybersecurity risks, safeguard sensitive data, and ensure the continuity of operations in an increasingly interconnected and digitalized world. Furthermore, blockchain's potential extends beyond cybersecurity to encompass broader applications such as supply chain transparency, digital identity management, and smart contracts, all of which stand to revolutionise the way businesses operate and collaborate in the digital age. However, despite its promise, the widespread adoption of blockchain technology in Industry 4.0 also presents challenges and complexities that must be addressed to realise its full potential. Chief among these is the scalability and performance limitations inherent in blockchain networks, which may hinder their ability to handle the high transaction volumes and processing speeds required in industrial settings. Interoperability issues between different blockchain platforms and legacy systems may pose obstacles to seamless integration and adoption.

Furthermore, concerns related to regulatory compliance, data privacy, and governance may require careful consideration and adaptation of existing frameworks to accommodate the unique characteristics of blockchain technology. Nevertheless, the transformative potential of blockchain in securing the future of Industry 4.0 is undeniable. By providing a decentralised, transparent, and tamper-proof foundation for cybersecurity, blockchain empowers enterprises to navigate the complexities of the digital landscape with confidence and resilience. As Industry 4.0 continues to evolve, blockchain stands as a beacon guiding the way towards a future where digital assets and transactions are safeguarded, and trust is assured in an increasingly interconnected and digitised world.

5.6 Beyond Transactions: Exploring Blockchain Applications in Industry 4.0

While blockchain is often associated with financial transactions and cryptocurrencies, its potential applications extend far beyond traditional finance. In Industry 4.0, blockchain is being explored for diverse use cases such as supply chain management, digital identity, intellectual property protection, and decentralised finance (DeFi). In supply chain management, blockchain enables end-to-end visibility and traceability of goods, facilitating compliance with regulatory requirements, ethical sourcing practices, and sustainability initiatives. In digital identity, blockchain offers a secure and decentralised framework for managing identity credentials, enabling individuals to control access to their personal data and authenticate themselves without relying on centralised authorities. In intellectual property protection, blockchain enables timestamping and digital fingerprinting of creative works, providing immutable proof of ownership and authenticity. In decentralised finance (DeFi), blockchain enables peer-to-peer lending, borrowing, and trading of digital assets without the need for intermediaries, democratising access to financial services and fostering financial inclusion. By exploring blockchain applications beyond transactions, Industry 4.0 can unlock new opportunities for innovation, efficiency, and resilience in a rapidly evolving digital economy.

5.6.1 From Smart Contracts to Supply Chains

The evolution of blockchain technology within the landscape of Industry 4.0 marks a transformative journey characterised by a profound shift from basic transactional use cases to more intricate and integrated applications, heralding a new era of innovation and efficiency in business processes. Initially propelled into the spotlight by cryptocurrencies such as Bitcoin and Ethereum, blockchain garnered widespread attention and adoption through its ground-breaking feature of smart contracts.

These self-executing contracts, encoded directly onto the blockchain, revolutionised the way agreements are executed by automating processes based on predefined conditions, thereby imbuing transactions with unparalleled automation, transparency, and efficiency. Smart contracts found application across diverse domains, from procurement and logistics to payments and beyond, streamlining operations and mitigating the need for intermediaries. However, the true potential of blockchain technology transcends its transactional origins, as evidenced by its burgeoning integration into supply chain management—a domain ripe for disruption in the era of Industry 4.0. Leveraging the foundation laid by smart contracts, blockchain is now revolutionising supply chain management by offering end-to-end visibility and traceability of goods, mitigating the risk of counterfeiting and fraud, and optimising supply chain operations from procurement to delivery. By deploying blockchain solutions, businesses can track the journey of products across the entire supply chain in real-time, recording every transaction and transfer of ownership on an immutable ledger, thus ensuring transparency and accountability at every stage. This heightened transparency not only fosters trust among stakeholders but also enables swift identification and mitigation of issues such as product recalls, quality control discrepancies, and logistical bottlenecks, thereby enhancing operational efficiency and customer satisfaction.

Moreover, blockchain's inherent security features, such as cryptographic hashing and decentralised consensus mechanisms, fortify supply chains against unauthorised alterations and cyber threats, safeguarding sensitive data and intellectual property.

Additionally, blockchain facilitates the implementation of smart contracts within supply chain management, automating processes such as procurement, inventory management, and payments based on predefined conditions and triggering events.

This automation not only reduces administrative overheads and transaction costs but also accelerates the pace of operations, enabling businesses to respond swiftly to changing market dynamics and customer demands. Furthermore, blockchain's immutable nature and decentralised architecture make it an ideal tool for combating counterfeiting and fraud—a pervasive issue plaguing supply chains worldwide. By recording the provenance and authenticity of goods on the blockchain, businesses can verify the integrity of products, mitigate the risk of counterfeit goods entering the market, and protect their brand reputation.

As blockchain continues to evolve and mature, its integration into supply chain management represents a paradigm shift in the way businesses collaborate, transact, and operate in the digital age. By leveraging the transformative power of blockchain—from its inception in smart contracts to its current application in supply chains—enterprises can unlock new avenues for innovation, efficiency, and competitiveness in Industry 4.0. As the journey of blockchain's evolution unfolds, businesses must seize the opportunity to embrace this disruptive technology and harness its potential to reshape the future of supply chain management and beyond.

5.7 Data Integrity Revolution: How Blockchain is Reshaping Industry 4.0

Data integrity is a critical concern in Industry 4.0, where the volume, variety, and velocity of data are increasing exponentially. Traditional data management systems are often centralised, siloes, and vulnerable to data breaches, tampering, and corruption. Blockchain technology addresses these challenges by providing a decentralised and immutable ledger that ensures the integrity and authenticity of data through cryptographic verification and consensus mechanisms. In Industry 4.0, blockchain is being leveraged for various data management use cases such as data provenance, data sharing, and data monetization.

By recording data transactions on a tamper-proof blockchain ledger, organisations can establish a single source of truth, enhance data integrity and trustworthiness, and enable secure and transparent data sharing among stakeholders. By revolutionising data integrity through blockchain, Industry 4.0 can unlock new opportunities for data-driven innovation, decision-making, and value creation in a hyper connected and data-driven world.

5.8 Empowering Collaboration: Blockchain's Role in Industry 4.0 Ecosystems

Collaboration is a key driver of innovation and competitiveness in Industry 4.0, where interconnected ecosystems of suppliers, partners, and customers collaborate to co-create value. However, traditional collaboration models are often centralised, fragmented, and inefficient, leading to trust issues, data silos, and coordination challenges. Blockchain technology offers a decentralised and transparent framework for empowering collaboration among diverse stakeholders in Industry 4.0 ecosystems. By providing a shared and tamper-proof ledger, blockchain enables secure and transparent data sharing, automated execution of smart contracts, and real-time tracking of transactions and assets. In supply chain management, blockchain facilitates collaboration among suppliers, manufacturers, and logistics providers by providing end-to-end visibility and traceability of goods, reducing the risk of disputes and delays. Similarly, in digital ecosystems, blockchain enables decentralised governance and incentive mechanisms, fostering greater participation, innovation, and value creation.

5.9 The Expansion of Blockchain Applications in Industry 4.0

Beyond its initial applications in finance and supply chain management, blockchain technology is finding new use cases and applications across various industries within the realm of Industry 4.0. One such area is digital identity management. Traditional identity management systems are often centralised and prone to security breaches, leading to identity theft and fraud. Blockchain technology offers a decentralised approach to managing digital identities, where individuals have control over their personal data and can selectively share it with trusted parties. This not only enhances privacy and security but also streamlines processes such as user authentication and Know Your Customer (KYC) compliance.

Another burgeoning application of blockchain in Industry 4.0 is in the realm of decentralised autonomous organisations (DAOs). DAOs are organisations that operate without centralised control and are governed by smart contracts and consensus mechanisms on the blockchain. These organisations enable transparent and democratic decision-making processes, where members collectively determine the direction and operations of the organisation. DAOs have the potential to revolutionise traditional organisational structures, fostering greater transparency, accountability, and efficiency.

Furthermore, blockchain technology is being explored for its potential to revolutionise the healthcare industry in the context of Industry 4.0. Healthcare data is highly sensitive and often fragmented across various stakeholders, making it challenging to access and share securely. Blockchain offers a solution by providing a secure and transparent framework for storing and sharing medical records, enabling seamless interoperability among healthcare providers while ensuring patient privacy and data security. Additionally, blockchain can facilitate the tracking and authentication of pharmaceuticals throughout the supply chain, reducing the risk of counterfeit drugs and improving patient safety.

Moreover, blockchain technology is poised to transform the energy sector in the era of Industry 4.0. The energy industry is undergoing a paradigm shift towards decentralised and renewable energy sources, leading to complex energy trading and management processes. Blockchain can streamline energy transactions by enabling peer-to-peer energy trading networks, where consumers can buy and sell excess energy directly to each other without the need for intermediaries. This not only promotes energy efficiency and sustainability but also empowers consumers to have greater control over their energy usage and costs.

Additionally, blockchain technology is revolutionising the field of intellectual property (IP) management in Industry 4.0. Intellectual property rights are critical for incentivizing innovation and creativity, yet traditional IP management systems are often cumbersome and prone to disputes. Blockchain offers a transparent and immutable ledger for recording and managing IP rights, enabling creators to timestamp their works and establish ownership on a tamper-proof platform. This enhances trust and transparency in the IP ecosystem, facilitating licensing, royalties, and collaboration among creators, innovators, and rights holders.

In conclusion, the trajectory of blockchain technology presents a profound paradigm shift poised to revolutionise the very essence of Industry 4.0. As a secure, transparent, and decentralised framework, blockchain stands as a cornerstone of transformation across various facets of manufacturing and beyond, heralding an era of unparalleled innovation and collaboration. From the realms of supply chain management to the intricacies of digital identity, from the evolution of decentralised finance to the intricacies of healthcare, blockchain's impact permeates diverse industries, unlocking novel opportunities for efficiency, collaboration, and progress. Its immutable ledger and cryptographic protocols lay the foundation for trust and transparency, transcending traditional limitations and reshaping the way businesses operate and interact. As we delve deeper into the potential of blockchain within the context of Industry 4.0, it becomes increasingly evident that its transformative power extends far beyond mere technological advancement. It holds the promise of fundamentally reshaping the very fabric of how we conceive, create, and collaborate in the Fourth Industrial Revolution and beyond. By embracing blockchain's potential, industries can foster a new era of decentralised governance, transparent transactions, and collaborative ecosystems, driving forward the relentless march of progress towards a future characterised by innovation, efficiency, and inclusivity. As we stand on the precipice of unprecedented change, the transformative impact of blockchain technology serves as a beacon of hope, guiding us towards a future where boundaries are blurred, opportunities are limitless, and collaboration knows no bounds. In this era of perpetual evolution, blockchain emerges as a catalyst for change, reshaping the very foundation of Industry 4.0 and laying the groundwork for a future defined by innovation, collaboration, and prosperity.

Key Notes:

Topic	Keynotes
Blockchain Fundamentals	Decentralization, immutability, and cryptographic security
	Functioning as a distributed ledger for ensuring trust and transparency in transactions
Applications in Industry 4.0	Use cases such as supply chain management, data integrity, and cybersecurity
	Potential impact on traditional industrial practices and business models

Topic	Keynotes
Evolution of Blockchain	Historical development from Bitcoin to current iterations in Industry 4.0
	Advancements and challenges associated with the evolution of blockchain technology over time
Role in Trust Building	Enhancing trust and transparency in transactions and data management processes
	Implications for establishing trust in complex industrial ecosystems
Challenges and Opportunities	Barriers to adoption (e.g., scalability, regulatory concerns)
	Opportunities for innovation, efficiency, and collaboration in the digital age

Trivia Challenge Industry 4.0

Welcome to the Industry 4.0 Trivia Challenge! Get ready to test your knowledge of the wacky world of Industry 4.0 with these hilarious multiple-choice questions. Each question is paired with some side-splitting facts and figures about Industry 4.0. Let us dive in and have a giggle!

1) What does Blockchain bring to Industry 4.0 like a trendy accessory?

    A) A moustache and monocle for every factory worker.

    B) Decentralized innovation and trust-building magic.

    C) A pet rock named Satoshi Nakamoto.

    D) A funky blockchain dance party in the smart factory.

2) How does Blockchain reshape Industry 4.0 according to the chapter?

    A) By transforming factories into secret lairs for digital superheroes.

    B) By fostering decentralized innovation and trust within industrial processes.

    C) By turning assembly lines into blockchain-themed rollercoasters.

    D) By replacing conveyor belts with blockchain-powered magic carpets.

3) What does Blockchain do to innovation, according to the chapter?

    A) Bottles it up and sells it in limited edition NFTs.

    B) Spreads it like wildfire across a decentralized network.

    C) Keeps it under lock and key like a digital dragon hoarding treasure.

    D) Sends it on vacation to a virtual reality beach resort.

4) How does Blockchain build trust within industrial processes?

A) By hiring a team of blockchain-certified trust-builders.

B) By creating an immutable ledger that everyone can trust.

C) By hosting trust-building workshops for factory workers.

D) By bribing scepticism with digital coins.

5) What is one implication of Blockchain for data integrity in Industry 4.0?

A) It ensures data integrity through its tamper-proof nature.

B) It transforms data into digital butterflies that flutter freely.

C) It encrypts data with secret blockchain handshakes.

D) It replaces data integrity with digital high-fives.

6) How does Blockchain empower collaboration within Industry 4.0 ecosystems?

A) By hosting blockchain-themed tea parties for industry giants.

B) By providing a secure and transparent platform for collaboration.

C) By creating digital avatars for virtual teamwork.

D) By outsourcing collaboration to blockchain-powered robots.

7) What does Blockchain bring to the table in terms of Industry 4.0 applications?

A) A buffet of possibilities, from supply chain management to smart contracts.

B) A magic wand that turns factories into blockchain-powered castles.

C) A digital Swiss army knife for solving industrial puzzles.

D) A blockchain-powered time machine that transports factories to the future.

8) How does Blockchain revolutionize trust-building in industrial processes?

A) By hiring a team of blockchain-certified trust-builders.

B) By eliminating the need for intermediaries and creating trust through transparency.

C) By replacing factory workers with trusty blockchain-powered robots.

D) By casting a trust spell over the factory floor with blockchain magic.

9) What metaphor does the chapter describe Blockchain's impact on Industry 4.0?

A) Like adding rocket fuel to a factory's innovation engine.

B) Like laying down the tracks for a trust-powered industrial revolution.

C) Like putting a digital genie in every factory's lamp.

D) Like sprinkling blockchain fairy dust over outdated industrial practices.

10) What does Blockchain bring to the party in terms of trust?

A) A blockchain-powered lie detector for factory workers.

B) A trusty ledger that everyone can rely on.

C) A blockchain-themed trust fall exercise for team-building.

D) A digital handshake for sealing industrial deals.

Summary:

The learning objectives focus on providing a comprehensive understanding of blockchain technology and its applications in the context of Industry 4.0. Participants will gain insights into the fundamental principles of blockchain, its evolution over time, and its potential to reshape industrial practices. Additionally, the objectives aim to foster discussions on the role of blockchain in building trust, addressing challenges, and unlocking opportunities for innovation and collaboration in the Fourth Industrial Revolution.

Abbreviations

Abbreviations	Terms
BF	Blockchain Fundamentals
AI4.0	Applications in Industry 4.0
EoB	Evolution of Blockchain
RTB	Role in Trust Building
CAO	Challenges and Opportunities

Chapter 6

Robotics in the Era of Industry 4.0

Learning objectives:

- Understanding Industry 4.0 and Robotics Integration
- Fundamental Principles of Robotics
- Hardware Components in Robotics
- Advances in Robotics Technology
- Applications of Robotics in Industry 4.0
- Robotics Integration
- Challenges and Considerations in Robotics Integration
- Future Trends in Robotics for Industry 4.0

6.0 Basic Principles of Robotics

The Fourth Industrial Revolution and the Role of Robotics

The advent of Industry 4.0 marks a revolutionary shift in manufacturing and industrial processes, fundamentally changing how technology is integrated into daily life and work. This overview establishes context for understanding the symbiotic relationship between Industry 4.0 and robotics by exploring core definitions and how robotics enables this transformative industrial landscape.

Central to Industry 4.0 is robotics' transformative influence in shaping future industrial operations. Robotics within Industry 4.0 is characterised by integrating advanced robotic systems with intelligent technologies, enabling automation, autonomy, and data-driven decision making. These robotic systems go beyond traditional mechanisation by incorporating artificial intelligence, machine learning, and sensors to interact with their environment in real-time, make autonomous decisions, and seamlessly collaborate with human workers. Robotics' role spans sectors such as smart manufacturing, logistics, healthcare, and more—driving innovation and efficiency previously considered impossible. As the discussion continues, it is evident that robotics is not merely a tool within Industry 4.0 but rather a transformative force redefining the very nature of industrial processes and what can be achieved.

The foundations of robotics for Industry 4.0 extend beyond simple mechanisation. Robotics encompasses sophisticated principles and hardware that reimagine how machines perceive their surroundings and collaborate with people.

6.0.1 Automation and Autonomy

Automation refers to the use of machines to perform tasks without human intervention. In Industry 4.0, robots are capable of more than just automation; they can adapt and optimise their actions based on real-time data, making them intelligent entities.

Autonomy takes the concept of automation a step further by providing robots with decision-making abilities. In Industry 4.0, robots are designed for autonomy, enabling them to analyse data, make independent decisions, and carry out tasks without constant human oversight.

6.0.2 Human-Robot Collaboration

Coexistence in Shared Work Environments: Unlike traditional manufacturing settings where robots operate in isolated areas, Industry 4.0 emphasises human-robot collaboration in shared workspaces. This cooperative approach involves robots and human workers coexisting, sharing responsibilities, and working seamlessly together.

Collaborative Robots (Cobots): Industry 4.0 introduces cobots, robots specifically engineered to collaborate with humans. These robots can modify their behaviour based on human actions, ensuring a safe and productive work environment.

6.1 Hardware Components

Sensors and Actuators

Sensors: Sensors are the sensory organs of robotic systems, providing them with the ability to perceive and interact with their environment. In Industry 4.0, robots are equipped with a diverse range of sensors, including cameras, lidar, accelerometers, and more. These sensors enable robots to capture data about their surroundings, identify objects, and make informed decisions.

Actuators: Actuators are the muscles of robotic systems, responsible for executing physical actions based on the decisions made by the robot. In Industry 4.0, advanced actuators enable robots to perform intricate and precise movements. Electric motors, pneumatic systems, and hydraulics are common actuators that contribute to the agility and versatility of modern robotic systems.

Understanding these foundational principles and hardware components is crucial for appreciating the depth of robotics in Industry 4.0. The combination of automation, autonomy, and collaborative capabilities, coupled with advanced sensors and actuators, empowers robots to be integral contributors to smart, connected, and adaptable industrial processes. The subsequent chapters will delve into the diverse applications, benefits, and challenges associated with the implementation of robotics in the era of Industry 4.0.

Benefit of Robotic Process Automation							
Accuracy	Low Technical Barrier	Compliance	Non-Invasive Technology	Consistency	Reliability	Productivity	Improve Employee Morale
Extreme Accuracy and uniformity	No Programming Skills	Audit Trail	Easy to Configure	Highly Accuracy	Repeatability through Automation	Process Cycle Time Reduced	Safe Environment

Exhibit 6.1 Robotic Process Automation (RPA)

The Fourth Industrial Revolution, or Industry 4.0, is defined by the integrating of digital, physical, and biological technology. It transforms how we live, work, and interact with the world. Robotics is a key enabling technology for Industry 4.0.

The robotics industry has seen significant advances in recent years, driven by technological innovation and increasing demand for automation.

6.2 Some of the critical advances in robotics include

Collaborative robots (cobots): Cobots are robots that are designed to work safely alongside humans. They are typically smaller and more agile than traditional industrial robots and are easier to program and operate.

Autonomous mobile robots (AMRs): AMRs can navigate and operate without human intervention. They are often used in warehouses and other industrial facilities to transport materials and perform other tasks.

AI-powered robots: AI is being used to develop more intelligent and adaptable robots. These robots can learn from their experiences and make decisions in real-time.

Machine learning-enabled robots: Machine learning is being used to create robots that can learn from data and improve over time.

Deep learning-enabled robots: Deep learning creates robots capable of performing sophisticated tasks like image recognition and natural language processing.

These advances in robotics are enabling new and innovative applications in various industries. For example, robots are being used to: Automate tasks in manufacturing and logistics: Robots can perform a wide range of functions in manufacturing and logistics, such as welding, assembly, and picking and packing.

Assist with healthcare: Robots can be used to assist with a variety of healthcare tasks, such as patient care, surgery, and rehabilitation. It helps to improve the quality of care and reduce the workload on healthcare professionals.

Perform dangerous or repetitive tasks: It can perform complex or repetitive tasks, such as cleaning up hazardous waste sites or inspecting bridges and other infrastructure. It helps to protect workers and improve safety.

The advances in robotics in the era of Industry 4.0 are significantly impacting how we live and work. As robots become more intelligent, adaptable, and affordable, we can expect to see them being used in even more innovative and transformative ways in the years to come.

Industry 4.0, also known as the Fourth Industrial Revolution, represents the integration of digital technologies into various aspects of manufacturing and industry. Robotics plays a pivotal role in driving the transformative changes associated with Industry 4.0. In this context, the applications of robotics are diverse and impactful, contributing to increased efficiency, precision, and flexibility across various sectors. This section explores some key applications of robotics in Industry 4.0.

6.3 Smart Manufacturing

Automated Assembly Lines: In smart manufacturing, robotics is extensively used in automated assembly lines to streamline production processes. Robots equipped with advanced sensors and vision systems can perform intricate assembly tasks with high precision and speed. Collaborative robots (cobots) work alongside human workers, enhancing productivity and flexibility. The integration of machine learning algorithms enables robots to adapt to variations in production and make real-time adjustments, optimising the overall manufacturing process.

Robotic Process Automation (RPA): Robotic Process Automation involves the use of software robots to automate repetitive and rule-based tasks within business processes. In smart manufacturing, RPA is employed to automate administrative and data-related tasks, such as inventory management, order processing, and quality control. This not only reduces human error but also accelerates the decision-making process, leading to increased operational efficiency.

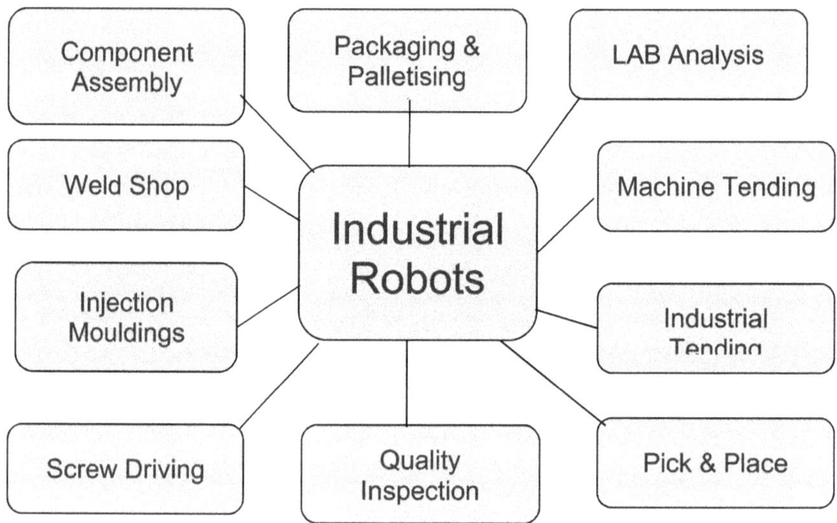

Exhibit 6.2 Industrial Robots Applications

6.4 Warehouse and Logistics

Automated Material Handling

Robotic technologies are transforming warehouse and logistics operations through automated material handling. Automated guided vehicles and robotic arms navigate warehouses using integrated sensors to efficiently and accurately pick, transport, and stock inventory. These robots can work collaboratively, coordinating movements to optimise material flow throughout the supply chain. Integration with Internet of Things devices provides real-time tracking and monitoring of inventory levels.

Autonomous Vehicles and Drones

Within Industry 4.0, autonomous vehicles and drones are utilised for logistics and transportation purposes. Autonomous trucks and drones equipped with robotic arms can autonomously load and unload shipments, contributing to efficiencies in supply chain operations. Additionally, these technologies reduce reliance on human labour for transportation activities, helping to mitigate labour shortages while enhancing safety across logistics networks.

6.5 Healthcare and Life Sciences

6.5.1 Surgical Robots

The healthcare sector has witnessed a revolution in surgical procedures with the introduction of robotic systems. Surgical robots, controlled by skilled surgeons, offer enhanced precision and dexterity in minimally invasive procedures. They provide 3D visualisation, allowing surgeons to perform complex surgeries with smaller incisions, reducing patient recovery time and improving overall outcomes.

Rehabilitation Robotics:

Industry 4.0 has brought about advancements in rehabilitation through the use of robotics. Exoskeletons and robotic rehabilitation devices assist individuals with mobility impairments in regaining motor functions. These devices, often connected to data analytics platforms, enable personalised rehabilitation programs, monitoring progress and adapting to the patient's evolving needs.

The applications of robotics in Industry 4.0 are diverse and transformative, impacting various sectors ranging from manufacturing and logistics to healthcare. As technology continues to advance, the integration of robotics is likely to expand, driving further innovation and efficiency across industries.

Benefits and Impacts of Robotics in Industry 4.0

The integration of robotics in Industry 4.0 brings about a myriad of benefits and transformative impacts, reshaping the landscape of various industries. From manufacturing to healthcare, these advancements contribute to increased efficiency, safety, and flexibility, while also addressing challenges associated with labour intensity. This section delves into the key benefits and impacts of incorporating robotics into Industry 4.0.

Increased Efficiency and Precision:

Speed and Productivity: Robotics significantly enhances the speed of tasks and overall productivity. Automated processes, whether in manufacturing or logistics, operate at a consistent pace, reducing cycle times and increasing throughput. Robots excel in repetitive tasks, enabling them to work continuously without fatigue, resulting in higher production rates.

Precision and Accuracy: Robots equipped with advanced sensors, vision systems, and machine learning algorithms demonstrate unparalleled precision in executing tasks. This is particularly crucial in manufacturing, where precision is paramount for quality control and assembly processes. The use of robotics minimises errors, leading to higher-quality outputs and reduced waste.

Reduction in Labour Intensity:

Labour Cost Savings: By automating routine and labour-intensive tasks, companies can achieve significant cost savings associated with human labour. This allows businesses to allocate resources more efficiently and invest in areas that require human expertise, creativity, and problem-solving skills.

Mitigation of Labour Shortages: Industries facing challenges related to labour shortages, such as logistics and manufacturing, can use robotics to fill the gaps. Robots can take on physically demanding or hazardous tasks, reducing reliance on a scarce labour pool and improving overall operational continuity.

Enhanced Safety:

Risk Mitigation: Robots are designed to handle dangerous tasks, such as working in hazardous environments or handling toxic materials. By delegating such tasks to robots, the risk of workplace accidents and injuries to human workers is significantly reduced, contributing to improved occupational safety.

Ergonomic Considerations: Collaborative robots (cobots) are designed to work alongside humans, assisting them in tasks that may pose ergonomic challenges. This collaborative approach not only enhances productivity but also minimises the risk of repetitive strain injuries and other musculoskeletal issues among human workers

Scalability and Flexibility:

Adaptability to Change: Robotics in Industry 4.0 allows for quick and efficient adaptation to changes in production demands. Automated systems can be reprogrammed or reconfigured to accommodate variations in product design or manufacturing requirements, providing a level of flexibility that traditional manufacturing processes may struggle to achieve.

Scalability of Operations: The scalability of robotic systems enables businesses to easily scale their operations in response to market demands. Whether it's increasing production volumes or diversifying product lines, robotic solutions can be scaled up or down to align with business needs, providing a competitive advantage in dynamic markets.

In summary, the integration of robotics in Industry 4.0 yields multifaceted benefits, ranging from improved efficiency and precision to enhanced safety and flexibility. As industries continue to embrace these technologies, the positive impacts on productivity, cost-effectiveness, and overall operational excellence are expected to become even more pronounced.

6.6 Challenges and Considerations in the Integration of Robotics in Industry 4.0

The adoption of robotics in Industry 4.0 brings about numerous advantages, but it requires addressing several challenges to ensure successful integration and sustained benefits across various industries.

Integration complexity can be an issue. Ensuring seamless communication between different devices, software platforms, and sensors through careful planning and standardised protocols is important for interoperability. As robotic systems gather and transmit data, safeguarding sensitive information from cyber threats and unauthorised access is paramount to protect intellectual property and ensure compliance with data protection regulations given increased connectivity and data exchange. The initial investment required for deployment, including hardware, software, and training, can be substantial, necessitating careful financial planning and consideration of long-term benefits, particularly for small and medium-sized enterprises.

Ethical considerations are also important. Automation may lead to job displacement, requiring prioritising reskilling and upskilling programs to prepare employees for new roles. For AI-powered systems making decisions and carrying out tasks, establishing clear accountability and transparency in how algorithms operate is crucial to build trust among stakeholders given potential lack of transparency. Addressing and mitigating bias is also important to ensure fair and just treatment and avoid discriminatory outcomes affecting individuals or groups disproportionately.

Workforce adaptation and training challenges must also be addressed. The integration of robotics requires developing skills in robotics programming, maintenance, and troubleshooting through comprehensive training programs to bridge skills gaps. Managing cultural and organisational change and fostering a culture of innovation and collaboration are also essential for a smooth transition to a more automated environment. Ensuring human worker safety, defining clear roles, and establishing effective communication channels are crucial for successful human-robot collaboration when implementing collaborative robots (Cobots).

In conclusion, while robotics integration offers benefits, proactively addressing technical integration complexities, ethical considerations, and workforce adaptation challenges through a holistic approach is required to navigate the rapidly evolving technological landscape successfully and ensure sustained benefits.

6.7 Future Trends in Robotics for Industry 4.0

As technology continues to advance, the future of robotics in Industry 4.0 holds exciting possibilities. Several emerging trends are shaping the evolution of robotic systems, offering new capabilities and enhancing their role in various industries. This section explores three key future trends in robotics for Industry 4.0.

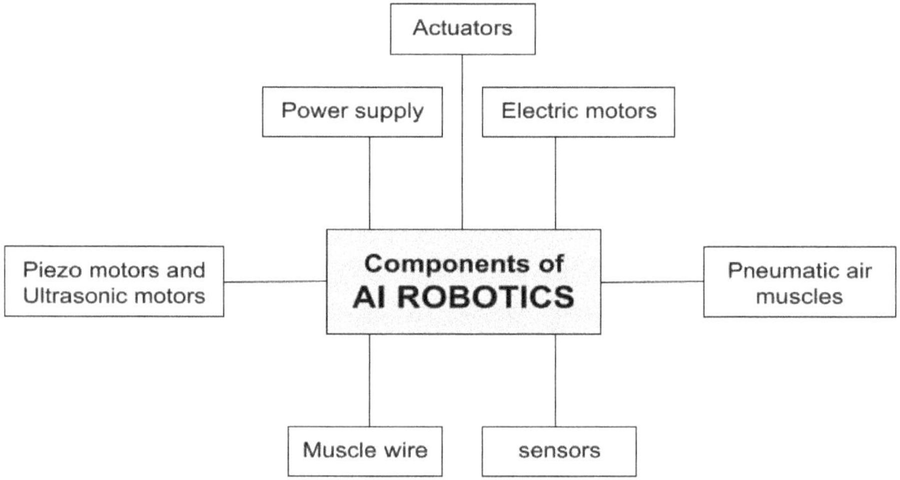

Exhibit 6.3 AI Robot Component

6.7.1 AI Integration for Intelligent Decision-Making:

Machine Learning and Adaptive Systems: The integration of artificial intelligence (AI) in robotics is evolving toward more sophisticated machine learning and adaptive systems. Future robotic systems will have the capability to continuously learn from data, enabling them to adapt to changing environments, optimise performance, and make intelligent decisions in real-time. This trend is particularly relevant in scenarios where complex decision-making processes are required, such as in autonomous vehicles, manufacturing, and logistics.

Cognitive Robotics: Cognitive robotics is an emerging field that aims to imbue robots with human-like cognitive abilities, including perception, reasoning, and problem-solving. By incorporating advanced AI techniques, future robotic systems will be better equipped to understand and interpret their surroundings, communicate effectively, and engage in complex tasks with a higher level of autonomy. This trend is expected to revolutionise industries such as healthcare, where robots can assist in more nuanced and interactive roles.

Swarm Robotics and Collaborative Systems:

Swarm Robotics: The concept of swarm robotics involves the coordination of multiple robots working collaboratively, similar to how a swarm of insects operates. Future trends in robotics envision the deployment of swarm robotics in scenarios where a collective, decentralised approach is advantageous. This can include applications in search and rescue, environmental monitoring, and distributed manufacturing processes, where a group of robots can accomplish tasks more efficiently through collaboration. Human-Robot Collaboration Advancements: The future will see advancements in human-robot collaboration, with a focus on seamless integration and cooperation between robots and human workers. Collaborative robots (cobots) will become more intuitive, capable of understanding human gestures and intentions. This trend aims to enhance productivity by leveraging the strengths of both humans and robots in a shared workspace, fostering a more adaptable and responsive manufacturing environment.

Biomechanical Enhancements in Robotics:

Soft Robotics and Bio-Inspired Designs: Future robotic systems are likely to incorporate soft robotics and bio-inspired designs, mimicking the flexibility and adaptability of natural organisms. Soft robotics enable robots to interact more safely with humans and delicate objects, expanding their use in environments that require a gentle touch. Bio-inspired designs draw inspiration from nature to create robots with enhanced agility and efficiency, leading to breakthroughs in fields such as medical robotics and search and rescue. Biomechanical Enhancements for Human Augmentation: The integration of biomechanical enhancements in robotics aims to augment human capabilities. Exoskeletons and wearable robotic devices will become more sophisticated, providing enhanced strength, agility, and endurance to individuals in various industries. This trend has significant implications for healthcare, manufacturing, and logistics, where human workers can benefit from increased physical capabilities without compromising safety.

The future trends in robotics for Industry 4.0 point toward a convergence of advanced technologies, including AI integration for intelligent decision-making, the emergence of swarm robotics and collaborative systems, and the development of biomechanical enhancements. These trends promise to revolutionise industries, unlocking new possibilities for efficiency, adaptability, and human-robot collaboration.

Here are some specific examples of how companies are using advances in robotics in the era of Industry 4.0 -

Logistic Companies like Amazon, flipkart, use robots in its warehouses to pick and pack orders. This has helped Amazon to improve efficiency and reduce costs.

Automotive Companies like Maruti, Honda, Toyota use robots in their factories to assemble their cars. This has helped them to produce cars more quickly and efficiently.

These are just a few examples of the many ways that advances in robotics are being used in the era of Industry 4.0. We expect to see even more innovative and transformative applications as robotics technology develops.

Keynotes

Topics	Keynotes
Industry 4.0 Paradigm Shift	Represents a transformative shift in manufacturing, integrating advanced technologies.
Role of Robotics	Central enabler of Industry 4.0, driving automation, autonomy, and data-driven decision-making.
Principles of Robotics	Automation, autonomy, and human-robot collaboration are foundational principles of robotics.
Hardware Components	Sensors and actuators enable robots to perceive, interact, and execute tasks in real-world environments.
Advances in Robotics	Recent advancements include collaborative robots, autonomous mobile robots, and AI-powered systems.
Applications Across Industries	Robotics revolutionizes various industries, from smart manufacturing and logistics to healthcare.
Benefits and Impacts	Integration yields increased efficiency, precision, safety, and flexibility across industries.
Challenges and Considerations	Integration complexity, ethical concerns, and workforce adaptation are key challenges.
Future Trends	Emerging trends include AI integration, cognitive robotics, and biomechanical enhancements.
Implications for Industry	Emphasizes the importance of embracing technological advancements for innovation and competitiveness.

Trivia Challenge: Industry 4.0

Welcome to the Industry 4.0 Trivia Challenge! Get ready to test your knowledge of the wacky world of Industry 4.0 with these hilarious multiple-choice questions. Each question is paired with some side-splitting facts and figures about Industry 4.0. Let us dive in and have a giggle!

1. What do readers gain upon completing this section?

A) A comprehensive understanding of how to build a robot army for world domination.

B) A profound grasp of Robotics' role in Industry 4.0, minus the robot uprising.

C) A recipe for robotic spaghetti—because why not?

D) A crash course in tap-dancing with robots.

2. What are the fundamental principles explored in this section?

A) The art of robot karaoke and synchronized dancing.

B) The delicate balance of human emotions and robot algorithms.

C) Automation, autonomy, and human-robot collaboration.

D) How to teach a robot to appreciate fine art.

3. What do readers dissect in this section?

A) The mysteries of the universe.

B) The essential building blocks of robotic systems.

C) The perfect recipe for robotic pancakes.

D) The latest gossip among robot celebrities.

4. What significant advancements in robotics are explored?

A) The invention of the robotic sock sorter.

B) The application of robots in making the perfect cup of coffee.

C) Robotics in smart manufacturing, warehouse and logistics, and healthcare and life sciences.

D) Teaching robots to appreciate the beauty of sunsets.

5. What challenges are addressed in integrating robotics into Industry 4.0?

A) Convincing robots not to take over the world.

B) The difficulties of finding robot-friendly office spaces.

C) Navigating the hurdles of incorporating robotics seamlessly into industrial processes.

D) Getting robots to understand human humour.

6. How does this section explore the transformative impact of robotics?

A) By creating a robot dance-off competition.

B) Through examining its applications in various industries within Industry 4.0.

C) By teaching robots to write poetry.

D) By inventing a robot language translation device.

7. What future trends in robotics are discussed in this section?

A) Robots learning to bake cakes and cookies.

B) The rise of robot politicians in global governance.

C) Insights into the evolving landscape of robotics within Industry 4.0.

D) Robot vacations to tropical islands.

8. How does this section address the dynamics of human-robot collaboration?

A) By organizing team-building exercises for humans and robots.

B) Through exploring how humans and robots work together for enhanced productivity.

C) By teaching robots to understand human emotions better.

D) By creating a robot matchmaking service.

9. What insights do readers gain about robotics' role in smart manufacturing?

A) How robots are trained to become expert chefs.

B) The secrets behind robots' impeccable fashion sense.

C) The application of robotics in optimizing manufacturing processes.

D) How robots are revolutionizing the art of gardening.

10. What do readers learn about the hardware components of robotic systems?

A) How to build a robot from scratch using household items.

B) The best tools for giving robots a makeover.

C) The essential building blocks that make up robotic systems.

D) How robots accessorize to express their personalities.

Summary:

This Chapter delves into the intersection of robotics and Industry 4.0, providing an in-depth exploration of how robotics is reshaping modern manufacturing and industrial processes. It begins by establishing the context of Industry 4.0 as a transformative paradigm shift driven by the integration of advanced technologies, with robotics playing a central role. Robotics within Industry 4.0 is characterized by its ability to integrate with intelligent systems, enabling automation, autonomy, and data-driven decision-making. The chapter elucidates the fundamental principles of robotics, including automation, autonomy, and human-robot collaboration, highlighting how these principles underpin the advancements in robotic technology. It examines key hardware components such as sensors and actuators, elucidating their role in enabling robots to perceive and interact with their environment. Advances in robotics technology are discussed, including collaborative robots (cobots), autonomous mobile robots (AMRs), and AI-powered robots, showcasing their diverse applications across industries. Real-world examples illustrate how robotics is revolutionizing smart manufacturing, warehouse and logistics operations, healthcare, and life sciences. The chapter emphasizes the benefits and impacts of robotics integration in Industry 4.0, ranging from increased efficiency and precision to enhanced safety and flexibility. It also addresses the challenges and considerations associated with robotics adoption, including integration complexity, ethical concerns, and workforce adaptation. Future trends in robotics for Industry 4.0 are explored, including AI integration for intelligent decision-making, cognitive robotics, swarm robotics, and biomechanical enhancements. These trends promise to further revolutionize industries, unlocking new possibilities for efficiency, adaptability, and human-robot collaboration.

Abbreviations

Abbreviations	Terms
ML	Machine Learning
DL	Deep Learning
3D	Three-Dimensional
AMRs	Autonomous Mobile Robots
IP	Intellectual Property
GDPR	General Data Protection Regulation
SMEs	Small and Medium-sized Enterprises

Abbreviations	Terms
AI	Artificial Intelligence
HR	Human Resources
RPA	Robotic Process Automation
AI	Artificial Intelligence
AMR	Autonomous Mobile Robot
IoT	Internet of Things
Cobots	Collaborative Robots

Chapter 7

Cloud Computing in Industry 4.0

Learning Objectives:

- Exploring Cloud-Based Services
- Analysing Cloud Empowerment Across Industries
- Assessing the Transition to Smart Factories with Cloud Computing
- Exploring Cloud-Driven Innovation and Agility
- Understanding the Scalability, Flexibility, and Intelligence of Cloud Computing
- Analysing Virtualization, Connectivity, and Security in Cloud Computing
- Evaluating the Role of Cloud Computing in Industry Transformation

7.1 Significance of Cloud Computing

As humanity strides, confidently towards the precipice of Industry 4.0, the amalgamation of digital technologies is heralding a seismic shift in the landscape of manufacturing and industrial processes. Within this epochal transition, cloud computing emerges as a towering colossus, offering scalable, adaptive, and economically viable solutions to meet the ever-evolving demands of contemporary industries. This comprehensive exploration embarks on an exhaustive odyssey, plunging into the depths of the profound influence wielded by cloud computing within the realm of Industry 4.0. Through meticulous scrutiny, we peel back the layers to reveal the multifaceted applications, myriad advantages, formidable challenges, and sweeping ramifications encapsulated within the domain of cloud computing. As we traverse this intricate terrain, we unearth the transformative potential inherent in cloud computing, poised to redefine the fundamental paradigms governing industrial operations and innovation in the epoch of Industry 4.0.

The emergence of Industry 4.0 represents a watershed moment in human history, characterised by the fusion of cyber-physical systems, the Internet of Things (IoT), artificial intelligence (AI), and big data analytics. These synergistic forces propel manufacturing and industrial processes into a new era, marked by unprecedented levels of efficiency, flexibility, and connectivity. Central to this transformative journey is the ascendance of cloud computing, which serves as the linchpin enabling the seamless integration and orchestration of disparate technologies within the industrial ecosystem.

One of the most conspicuous manifestations of cloud computing's impact on Industry 4.0 lies in its ability to facilitate real-time data collection, analysis, and decision-making across the manufacturing value chain. By leveraging cloud-based platforms and services, enterprises can harness the power of big data analytics to extract actionable insights from vast repositories of sensor data generated by interconnected devices and machines. This enables predictive maintenance, quality control, and process optimization, thereby enhancing productivity, reducing downtime, and minimising costs.

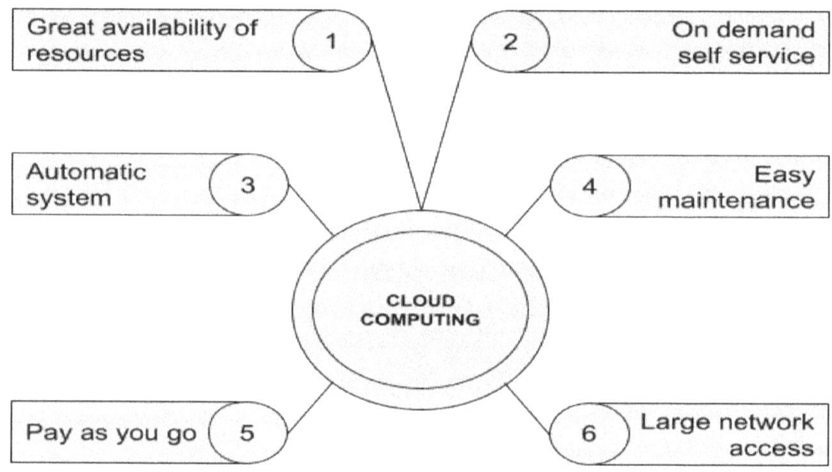

Exhibit 7.1 Cloud Services as a business models

Furthermore, cloud computing empowers manufacturers to embrace a more agile and adaptive approach to production through virtualization and simulation technologies. By migrating critical processes and workflows to cloud-based environments, companies can create digital twins of their physical assets, enabling them to conduct virtual experiments, simulate various scenarios, and iterate rapidly to optimise performance and mitigate risks. This not only accelerates product development cycles but also fosters innovation by enabling experimentation without the constraints of physical prototypes.

Moreover, cloud computing democratises access to advanced technologies and resources, particularly for small and medium-sized enterprises (SMEs) that may lack the financial means or technical expertise to deploy and maintain sophisticated IT infrastructure in-house.

Cloud-based services such as

- Software as a Service (SaaS),

- Platform as a Service (PaaS), and

- Infrastructure as a Service (IaaS)

offer SMEs a cost-effective and scalable alternative to traditional on-premises solutions, levelling the playing field and fostering greater competition and collaboration within the industry.

However, despite its myriad benefits, the widespread adoption of cloud computing in the context of Industry 4.0 also poses formidable challenges and concerns. Chief among these is the issue of data security and privacy, as the migration of sensitive industrial data to third-party cloud providers raises concerns about unauthorised access, data breaches, and regulatory compliance. Moreover, lots of third party cloud infrastructure service providers introduce new vulnerabilities and attack vectors, necessitating robust cybersecurity measures and risk mitigation strategies to safeguard critical assets and intellectual property.

Additionally, the increasing complexity and interdependence of cloud-based ecosystems necessitate greater interoperability and standardisation to ensure seamless integration and compatibility across disparate systems and platforms. This requires industry stakeholders to collaborate closely with standards bodies, regulatory agencies, and technology vendors to develop common frameworks, protocols, and best practices that promote interoperability, data portability, and vendor neutrality.

Furthermore, the pervasive nature of cloud computing engenders concerns about digital sovereignty and dependence on external providers, particularly in the context of national security, geopolitical tensions, and regulatory uncertainties. As governments and regulatory bodies grapple with issues of data localization, cross-border data flows, and jurisdictional disputes, manufacturers must navigate a complex regulatory landscape while balancing the benefits of global connectivity with the imperatives of data sovereignty and compliance.

Despite these challenges, the transformative potential of cloud computing in the era of Industry 4.0 remains undeniable. By harnessing the power of cloud-based technologies, manufacturers can unlock new levels of efficiency, innovation, and competitiveness, driving sustainable growth and prosperity in an increasingly digital and interconnected world.

However, realising this vision requires a concerted effort to address the technical, organisational, and regulatory challenges inherent in the adoption and integration of cloud computing within the industrial ecosystem. Only through collaborative innovation, strategic foresight, and a steadfast commitment to security and compliance can we fully harness the transformative potential of cloud computing to shape the future trajectory of manufacturing and industrial processes in the era of Industry 4.0.

7.2 Cloud Empowering Industry 4.0, Revolutionising Manufacturing and Beyond

Cloud computing stands as a foundational element of Industry 4.0, heralding a paradigm shift in traditional manufacturing practices and unlocking unprecedented levels of efficiency, adaptability, and innovation. By harnessing cloud-based infrastructure, organisations gain access to a rich array of computing resources on-demand, encompassing storage, processing power, and advanced analytics capabilities. This empowers manufacturers to optimise production processes, streamline supply chain operations, and foster collaboration among geographically dispersed teams. Moreover, cloud-based solutions seamlessly integrate with other Industry 4.0 technologies such as IoT, AI, and big data analytics, facilitating real-time insights and informed decision-making.

Cloud computing serves as the cornerstone upon which the edifice of Industry 4.0 is built, revolutionising manufacturing practices and extending its transformative influence beyond the factory floor to encompass diverse sectors such as healthcare, retail, education, and entertainment.

In healthcare, cloud-based solutions facilitate the storage and analysis of vast amounts of patient data, enabling personalised treatment plans and predictive analytics to improve patient outcomes.

In retail, cloud-powered platforms enhance customer engagement through personalised recommendations, seamless shopping experiences, and inventory management optimization.

In education, cloud-based learning management systems empower educators to deliver dynamic and interactive lessons, tailored to individual student needs, while facilitating collaboration and knowledge sharing among students and educators alike.

In entertainment, cloud gaming platforms enable gamers to access high-quality gaming experiences on-demand, without the need for expensive hardware or software installations.

Across these diverse domains, cloud computing serves as a catalyst for innovation, enabling organisations to adapt and thrive in the fast-paced, interconnected world of Industry 4.0. As we journey further into the digital age, the influence of cloud computing in revolutionising manufacturing and beyond will only continue to grow, driving unparalleled levels of efficiency, agility, and innovation across industries and reshaping the very fabric of our society.

7.3 Transitioning from Data Centres to Smart Factories: Cloud Computing in Industry 4.0

The transition from conventional data centres to cloud-based architectures serves as a linchpin of Industry 4.0 transformation, marking a pivotal shift in the landscape of manufacturing and industrial processes. Cloud computing empowers manufacturers to pivot from capital-intensive, on-premises IT infrastructure to scalable and cost-efficient cloud-based solutions, heralding a new era of agility and adaptability.

This paradigm shift enables companies to dynamically adjust computing resources based on demand, thereby reducing operational costs and enhancing overall efficiency. By leveraging the elasticity of cloud platforms, manufacturers can scale their operations up or down as needed, ensuring optimal resource allocation and minimising waste. Moreover, cloud platforms provide a secure and reliable environment for storing and processing vast amounts of data generated by IoT devices, sensors, and production equipment. This centralised repository of data serves as a valuable asset for manufacturers, enabling them to extract actionable insights, predict maintenance needs, and optimise operational workflows in real-time. Through advanced analytics and machine learning algorithms, cloud-based solutions empower manufacturers to identify patterns, trends, and anomalies within their data, facilitating proactive decision-making and continuous improvement. As manufacturers embrace the full potential of cloud computing, they gain unprecedented visibility and control over their operations, paving the way for enhanced productivity, competitiveness, and innovation in the Fourth Industrial Revolution and beyond.

7.4 Embracing Cloud Innovation: Enabling the Future of Industry 4.0

The widespread adoption of cloud computing catalyses the future of Industry 4.0 by endowing manufacturers with the agility and flexibility required to navigate evolving market dynamics and technological advancements, thus setting the stage for unprecedented levels of innovation, efficiency, and competitiveness. Cloud-based solutions serve as a catalyst for transformative change, fostering a culture of rapid prototyping and experimentation that enables companies to innovate and iterate at an accelerated pace, driving continuous improvement and value creation.

By leveraging the scalability and accessibility of cloud platforms, manufacturers can swiftly deploy and test new ideas, products, and services, allowing them to stay ahead of the curve in an ever-changing landscape. Moreover, cloud platforms facilitate seamless connectivity and collaboration across the extended enterprise, encompassing suppliers, partners, and customers, thus fostering a collaborative ecosystem where ideas can be shared, refined, and implemented collaboratively.

This culture of innovation and co-creation propels organisations towards greater levels of agility and responsiveness, enabling them to adapt and thrive amidst disruptive forces and emerging opportunities. Additionally, cloud computing empowers manufacturers to embrace emerging technologies such as edge computing, AI, and machine learning, further amplifying operational efficiency and competitiveness. By harnessing the power of AI-driven insights and predictive analytics, organisations can optimise production processes, anticipate customer needs, and identify opportunities for cost reduction and revenue growth.

Furthermore, edge computing enables real-time data processing and analysis at the edge of the network, reducing latency and enhancing responsiveness in critical applications such as predictive maintenance and quality control. As manufacturers continue to embrace the transformative potential of cloud computing, they will unlock new opportunities for growth, innovation, and value creation, thus shaping the future of Industry 4.0 and redefining the competitive landscape in the digital age.

7.5 Scalability and Flexibility: Cloud Computing's Role in Industry 4.0

The inherent scalability and flexibility of cloud computing represent pivotal advantages in the context of Industry 4.0, offering manufacturers a robust foundation upon which to navigate the dynamic and ever-evolving landscape of modern industry. Cloud-based solutions present a paradigm shift, providing virtually limitless scalability that empowers manufacturers to swiftly adjust computing resources to meet fluctuating production volumes or seasonal demand patterns, thus ensuring optimal resource allocation and minimising operational costs. This agility is particularly invaluable in industries characterised by dynamic operational environments, where the ability to scale up or down rapidly can mean the difference between success and stagnation.

Furthermore, cloud platforms offer the versatility to deploy and manage diverse workloads, ranging from batch processing to real-time analytics and high-performance computing, thereby enabling manufacturers to adapt their computing infrastructure to suit the specific requirements of different tasks and processes. This flexibility not only enhances operational efficiency but also fosters innovation and experimentation, as manufacturers can easily explore new technologies and approaches without being constrained by rigid infrastructure limitations. Moreover, the seamless integration of cloud-based solutions with existing IT systems and applications facilitates smooth migration and interoperability across hybrid environments, ensuring that manufacturers can leverage their existing investments while harnessing the power of the cloud to drive continuous improvement and innovation. As Industry 4.0 continues to evolve, the scalability and flexibility of cloud computing will play an increasingly central role in enabling manufacturers to stay competitive and responsive to changing market dynamics, thus shaping the future of manufacturing and industrial processes in the digital age and beyond.

7.6 Cloud-Driven Innovation: Transforming Industries in the Fourth Industrial Revolution

Cloud computing serves as a catalyst for innovation across industries in the Fourth Industrial Revolution, serving as a platform for experimentation, collaboration, and digital transformation. Manufacturers leverage cloud-based solutions to conceptualise, develop, and deploy innovative products, services, and business models that disrupt traditional markets and unlock new avenues for growth. From predictive maintenance and remote monitoring to digital twins and autonomous systems, cloud computing fuels the next wave of industrial innovation. Additionally, cloud platforms grant access to a diverse ecosystem of third-party services, APIs, and development tools, enabling manufacturers to expedite time-to-market and foster continuous innovation.

7.7 Virtualization and Connectivity: Cloud's Contribution to Industry 4.0

Virtualization and connectivity lie at the heart of Industry 4.0, with cloud computing playing a pivotal role in facilitating both. Cloud-based virtualization technologies empower manufacturers to abstract and pool computing resources, facilitating efficient resource utilisation and workload management. This capability enables companies to deploy virtualized environments for development, testing, and production, thus reducing infrastructure costs and enhancing scalability. Furthermore, cloud platforms provide the connectivity and interoperability necessary to integrate disparate systems and devices across the manufacturing ecosystem. This facilitates seamless data exchange, process orchestration, and collaboration, driving operational efficiency and agility.

The advent of cloud computing heralds a new era of agility in Industry 4.0 operations, serving as a pivotal linchpin for manufacturers navigating the complexities of the Fourth Industrial Revolution. In this dynamic landscape, characterised by rapid technological advancements and evolving market dynamics, agility emerges as a cornerstone of success, enabling enterprises to swiftly adapt to changing circumstances and meet the demands of an increasingly discerning customer base.

Cloud-based solutions, with their inherent flexibility and scalability, empower manufacturers to respond to these challenges with unprecedented speed and efficiency. By leveraging cloud computing, companies can deploy and scale applications and services with remarkable agility, reducing time-to-market and enhancing responsiveness to customer needs. The scalability of cloud platforms allows manufacturers to rapidly adjust computing resources in line with fluctuating demand, ensuring optimal performance and cost-efficiency even during peak periods.

Moreover, cloud platforms offer built-in automation and orchestration capabilities, streamlining and optimising business processes across the entire value chain. Through automated workflows and resource allocation, manufacturers can eliminate manual tasks and redundant processes, freeing up valuable time and resources to focus on innovation and value creation. Cloud computing democratises access to advanced technologies such as artificial intelligence, machine learning, and big data analytics, empowering manufacturers of all sizes to harness the transformative power of data-driven insights. By leveraging cloud-based analytics and predictive modelling, companies can gain deeper insights into customer preferences, market trends, and operational inefficiencies, enabling informed decision-making and proactive problem-solving. Additionally, cloud computing enables seamless collaboration and information sharing among stakeholders, regardless of geographical location or organisational boundaries. Cloud-based collaboration tools and platforms facilitate real-time communication and data exchange, fostering greater transparency, agility, and alignment across diverse teams and departments. This enhanced collaboration not only accelerates decision-making but also promotes innovation and knowledge sharing, driving continuous improvement and competitive advantage. Furthermore, cloud computing offers unparalleled resilience and reliability, with robust security measures and disaster recovery capabilities built into cloud infrastructure.

By entrusting critical business operations to reputable cloud service providers, manufacturers can mitigate the risks of data loss, cybersecurity threats, and downtime, ensuring business continuity even in the face of unforeseen disruptions.

In conclusion, the agility afforded by cloud computing is instrumental in shaping the future of Industry 4.0 operations, empowering manufacturers to thrive in an increasingly dynamic and competitive environment. By embracing cloud-based solutions, companies can unlock new opportunities for innovation, efficiency, and growth, positioning themselves at the forefront of the Fourth Industrial Revolution. As cloud computing continues to evolve and mature, its impact on industry operations will only intensify, driving greater agility, resilience, and success in the digital age.

7.8 Cloud-Based Intelligence: Enhancing Efficiency in Industry 4.0 Systems

Cloud-based intelligence drives efficiency in Industry 4.0 systems by furnishing manufacturers with real-time insights and actionable intelligence. Cloud platforms provide advanced analytics and machine learning capabilities, enabling companies to extract valuable insights from vast datasets. This facilitates predictive maintenance, quality optimization, and demand forecasting, thereby enhancing operational efficiency and curtailing downtime. Additionally, cloud-based AI services empower manufacturers to automate decision-making processes and optimise resource allocation, further amplifying efficiency and cost savings.

7.8.1 Data Liberation: Leveraging Cloud Computing for Industry 4.0 Insights:

Data liberation represents a pivotal benefit of cloud computing in Industry 4.0, enabling manufacturers to unlock the value of their data and derive actionable insights. Cloud platforms furnish scalable and cost-effective storage solutions that empower companies to capture, store, and analyse vast datasets generated by IoT devices, sensors, and production equipment. This enables manufacturers to gain visibility into their operations, identify optimization opportunities, and drive continuous improvement. Moreover, cloud-based data analytics tools provide robust visualisation and reporting capabilities, enabling companies to communicate insights and inform decision-making processes.

7.8.2 Towards Industry 4.0: The Paradigm Shift Enabled by Cloud Computing:

The adoption of cloud computing heralds a profound paradigm shift in manufacturing and industrial processes, marking a pivotal moment in the journey towards Industry 4.0 transformation. Cloud-based solutions stand as the cornerstone of this revolution, equipping manufacturers with scalable, flexible, and cost-effective computing resources that serve as the bedrock for driving innovation, efficiency, and competitiveness in the digital age. By embracing cloud computing, companies can expedite their digital transformation initiatives, streamlining operations and unlocking new growth opportunities that were previously inaccessible. The scalability of cloud-based solutions enables manufacturers to dynamically adjust their computing resources in response to fluctuating demand, ensuring optimal resource allocation and minimising operational costs. Moreover, the flexibility afforded by cloud platforms empowers organisations to adapt their computing infrastructure to meet the specific requirements of diverse tasks and processes, fostering agility and enabling rapid experimentation and innovation.

Additionally, cloud-based solutions facilitate seamless integration with existing IT systems and applications, facilitating smooth migration and interoperability across hybrid environments. This integration ensures that companies can leverage their existing investments while harnessing the power of the cloud to drive continuous improvement and innovation. As we journey deeper into the Fourth Industrial Revolution, the role of cloud computing will assume increasing centrality to the success of Industry 4.0 initiatives, reshaping the future of manufacturing and industrial processes in ways that were previously unimaginable. The transformative impact of cloud computing will extend far beyond individual organisations, driving systemic change across entire industries and revolutionising the way we conceive, create, and collaborate in the digital age and beyond. With cloud computing as its foundation, Industry 4.0 will continue to evolve, entering in an era of unprecedented innovation, efficiency, and connectivity that will redefine the very fabric of modern industry.

The integration of cloud computing into the fabric of Industry 4.0 marks a significant milestone in the ongoing evolution of manufacturing and industrial processes, heralding a new era of innovation, efficiency, and collaboration across diverse industries. Cloud-based solutions have emerged as powerful enablers, propelling organisations towards heightened levels of productivity, agility, and competitiveness. By harnessing the capabilities of cloud computing, manufacturers can revolutionise traditional practices, streamline operations, and unlock new avenues for growth and development. The transformative potential of cloud computing transcends mere optimization, it fundamentally reshapes the future landscape of manufacturing and beyond.

Cloud computing offers a host of benefits that resonate deeply with the imperatives of Industry 4.0. Its scalability, flexibility, and cost-effectiveness empower organisations to adapt swiftly to changing market dynamics and technological advancements. The dynamic nature of cloud-based resources enables companies to optimise resource allocation, minimise operational costs, and respond rapidly to fluctuating demand patterns. Moreover, the versatility of cloud platforms facilitates seamless integration with other Industry 4.0 technologies, such as IoT, AI, and big data analytics, fostering a holistic ecosystem of interconnected digital capabilities.

As we move forward into the Fourth Industrial Revolution, the centrality of cloud computing will only continue to grow, shaping the trajectory of manufacturing and industrial processes in profound ways. The transformative impact of cloud computing extends individual organisations, driving systemic change across entire industries and ecosystems. It fosters a culture of innovation, resilience, and sustainable growth, empowering organisations to thrive in an increasingly competitive and dynamic business landscape.

In the midst of this transformative journey, collaboration emerges as a cornerstone of success. Cloud computing facilitates seamless connectivity and collaboration across geographically dispersed teams, suppliers, and partners, fostering a culture of innovation and co-creation. By leveraging cloud-based platforms, organisations can transcend traditional boundaries, harnessing collective intelligence and expertise to drive continuous improvement and value creation.

Furthermore, the integration of cloud computing into Industry 4.0 brings with it a renewed focus on cybersecurity and data privacy. As organisations increasingly rely on cloud-based solutions to store and process sensitive data, ensuring robust security measures becomes paramount. Cloud providers invest heavily in state-of-the-art security protocols and compliance frameworks to safeguard data integrity and protect against cyber threats. By adopting best practices and leveraging advanced encryption technologies, organisations can mitigate risks and build trust among stakeholders, laying a solid foundation for sustainable growth and innovation.

Looking ahead, the future promises a landscape characterised by innovation, resilience, and sustainable growth, fuelled by the transformative potential of cloud computing. As organisations continue to embrace cloud-based solutions and leverage their capabilities to drive digital transformation initiatives, the possibilities are endless. From revolutionising supply chain management to enhancing customer experiences, cloud computing will continue to shape the way we conceive, create, and collaborate in the Fourth Industrial Revolution and beyond. In this era of rapid change and disruption, the centrality of cloud computing serves as a beacon of hope, guiding organisations towards a future defined by innovation, efficiency, and sustainable growth.

Key Notes:

Topics	Keynotes
Significance of Cloud Computing in Industry 4.0	- Cloud computing drives the transformation of manufacturing and industrial processes in the Fourth Industrial Revolution. - It enables real-time data collection, analysis, and decision-making, leading to enhanced productivity, reduced downtime, and cost savings.
Impact on Small and Medium-sized Enterprises (SMEs)	- Cloud-based services (SaaS, PaaS, IaaS) provide cost-effective and scalable alternatives to traditional on-premises solutions. - Challenges such as data security and interoperability need to be addressed for widespread adoption.
Empowerment Across Industries	- Cloud computing extends its transformative influence beyond manufacturing to sectors like healthcare, retail, education, and entertainment. - It enables personalized treatments, enhances customer engagement, facilitates dynamic learning experiences, and offers on-demand gaming.
Transition to Smart Factories	- Transition from conventional data centres to cloud-based architectures enables agility and adaptability in manufacturing.- Cloud platforms offer scalability, cost-efficiency, and data management benefits, empowering manufacturers to optimize operations and drive innovation.
Cloud-Driven Innovation and Agility	- Cloud computing fosters a culture of innovation, agility, and collaboration, enabling rapid prototyping, experimentation, and integration with emerging technologies. - It facilitates swift deployment and scaling of applications and services, streamlines workflows, and democratizes access to advanced technologies.

Scalability, Flexibility, and Intelligence	- Cloud computing offers inherent scalability and flexibility, allowing dynamic adjustment of computing resources and adaptation of infrastructure to specific requirements. - Cloud-based intelligence enhances operational efficiency through real-time insights, predictive maintenance, and demand forecasting.
Virtualization, Connectivity, and Security	- Cloud-based virtualization and connectivity optimize resource utilization, facilitate seamless data exchange, and enhance interoperability across the manufacturing ecosystem.- Robust security measures are essential to mitigate risks associated with data breaches and ensure compliance with regulatory requirements.
Role in Industry Transformation	- Cloud computing fundamentally reshapes manufacturing and industrial processes, driving continuous improvement, innovation, and competitiveness.- It serves as a catalyst for systemic change, enabling organizations to thrive in the digital age through collaboration, experimentation, and resilience.

Trivia Challenge: Industry 4.0

Welcome to the Industry 4.0 Trivia Challenge! Get ready to test your knowledge of the wacky world of Industry 4.0 with these hilarious multiple-choice questions. Each question is paired with some side-splitting facts and figures about Industry 4.0. Let us dive in and have a giggle!

1) What does Cloud Computing bring to Industry 4.0 like a superhero sidekick?

A) A cloud-shaped cape for every factory worker.

B) Scalability, flexibility, and agility to the rescue!

C) A cloud-powered ray gun for zapping inefficiencies.

D) A fog machine for cloud-themed parties in smart factories.

2) How does Cloud Computing empower Industry 4.0 initiatives, according to the chapter?

A) By summoning thunderstorms of innovation.

B) By providing scalable cloud-based services.

C) By transforming factories into fluffy cloud kingdoms.

D) By replacing factory floors with cloud-shaped trampolines.

3) **What does Cloud Computing do to smart factories in terms of innovation?**

   A) Turns them into cloud-powered amusement parks.

   B) Transforms them from data centres to smart factories.

   C) Hides them in the clouds like a digital game of hide-and-seek.

   D) Upgrades them with cloud-powered jetpacks for factory workers.

4) **According to the chapter, how does Cloud Computing shape the future of Industry 4.0?**

   A) By predicting factory futures with cloud-powered crystal balls.

   B) By driving innovation through cloud-based solutions.

   C) By hosting cloud-themed parties for industry giants.

   D) By clouding the minds of factory workers with futuristic dreams.

5) **What adjective describes learners' feelings about Cloud Computing in Industry 4.0?**

   A) Thunderous.

   B) Optimistic.

   C) Cloudy with a chance of efficiency.

   D) Fluffy.

6) **What feature of Cloud Computing makes learners optimistic about Industry 4.0?**

   A) Scalability, flexibility, and agility.

   B) Cloud-shaped cookies for factory workers.

   C) Cloud-based karaoke machines for factory parties.

   D) Cloud-powered teleporters for instant factory upgrades.

7) **How does Cloud Computing contribute to enhancing efficiency in Industry 4.0 systems?**

   A) By organizing cloud-themed treasure hunts for factory workers.

   B) By providing scalable, flexible, and agile solutions.

   C) By replacing factory floors with fluffy clouds for comfortable standing.

   D) By raining down efficiency from the cloud-filled skies.

8) **What does Cloud Computing bring to the table regarding Industry 4.0 innovation?**

   A) Scalability, flexibility, and agility.

   B) A cloud-shaped cake for every factory celebration.

   C) A cloud-powered crystal ball for predicting industry trends.

   D) Cloud-themed superhero costumes for factory workers.

9) **How does Cloud Computing transform data centres into smart factories?**

   A) By sprinkling them with magical cloud dust.

   B) By providing cloud-based services and solutions.

   C) By replacing factory workers with cloud-powered robots.

   D) By connecting data centres to the internet via fluffy clouds.

10) **What metaphor does the chapter use to describe Cloud Computing's impact on Industry 4.0?**

   A) Like a thunderstorm of innovation.

   B) Like a fluffy cloud lifting the industry to new heights.

   C) Like a cloud-powered rocket launching factories into the future.

   D) Like a digital rain shower refreshing outdated industrial practices.

Summary:

This Chapter explores the profound impact of cloud computing on Industry 4.0, showcasing its role in enabling scalable, adaptive solutions for modern manufacturing and industrial processes. It emphasizes how cloud-based platforms empower organizations to harness the power of real-time data analytics, agile production methods, and collaborative workflows. Despite challenges such as data security concerns and interoperability issues, cloud computing continues to revolutionize various sectors, including healthcare, retail, education, and entertainment. The chapter underscores the significance of cloud-driven innovation in driving efficiency, agility, and competitiveness across industries. By leveraging cloud-based intelligence and embracing digital transformation initiatives, organizations can unlock new opportunities for growth and development in the dynamic landscape of Industry 4.0.

Abbreviation

Abbreviation	Terms
API	Application Programming Interface
ML	Machine Learning
IT	Information Technology
GDPR	General Data Protection Regulation
AI-driven	Artificial Intelligence-driven
BPM	Business Process Management
DR	Disaster Recovery
IoT	Internet of Things
SaaS	Software as a Service
PaaS	Platform as a Service
IaaS	Infrastructure as a Service
SMEs	Small and Medium-sized Enterprises
AI	Artificial Intelligence
IIoT	Industrial Internet of Things
IP	Intellectual Property

Chapter - 8

Edge Computing in Industry 4.0

Learning Objective:

- Understand the concept of edge computing and its significance within the context of Industry 4.0.

- Explore the ways in which edge computing revolutionizes manufacturing and industrial operations.

- Examine the applications of edge computing, including real-time decision-making, proactive maintenance, and adaptive process adjustments.

- Analyse the advantages of edge computing, such as mitigating latency, minimizing bandwidth usage, and enhancing security measures.

- Discuss the challenges inherent in the adoption and integration of edge computing, including interoperability issues, security risks, and data management complexities. Gain insights into the implications of edge computing for the future of smart manufacturing and beyond.

- Recognize the transformative potential of edge computing in driving efficiency, agility, and innovation in the digital age.

8.1 Significance of Edge Computing

In the epoch defined by the dawn of Industry 4.0, the convergence of digital technologies is catalysing a profound metamorphosis in the landscape of manufacturing and industrial operations with unprecedented rapidity. Within this transformative milieu, edge computing emerges as a linchpin, poised to revolutionise the very fabric of industrial processes by bringing intelligence closer to the point of action and empowering real-time decision-making capabilities. This expansive inquiry embarks on an exhaustive odyssey, plunging into the depths of the profound influence wielded by edge computing within the realm of Industry 4.0. Through meticulous scrutiny, we unravel the multifaceted applications, myriad advantages, formidable challenges, and sweeping ramifications encapsulated within the domain of edge computing. As we navigate this complex landscape, we uncover the revolutionary potential of edge computing, which is ready to upend the core ideas guiding smart manufacturing and beyond.

At the heart of Industry 4.0 lies the seamless integration of cyber-physical systems, the Internet of Things (IoT), artificial intelligence (AI), and big data analytics, providing connectivity, automation, and intelligence in manufacturing and industrial processes. Central to this transformative journey is the advent of edge computing, which represents a paradigm shift in the way data is processed, analysed, and acted upon within industrial environments.

By bringing intelligence closer to the point of action, edge computing enables manufacturers to unlock new levels of efficiency, agility, and innovation, driving sustainable growth and competitiveness in an increasingly digital and interconnected world. However, realising this vision requires a concerted effort to address the

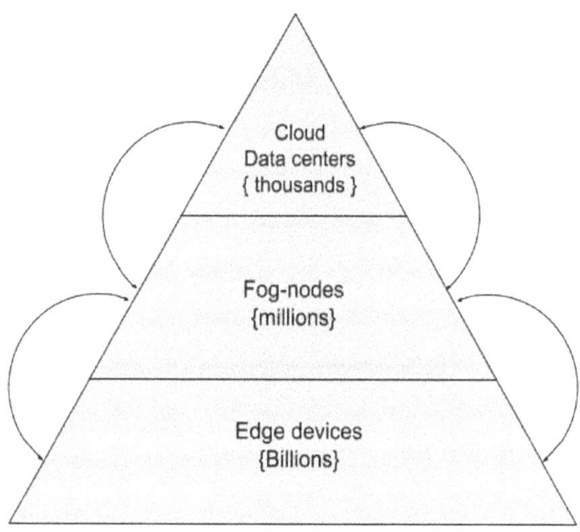

Exhibit 8.1 Type of Computing

Technical, organisational, and regulatory challenges inherent in the adoption and integration of edge computing within the industrial ecosystem. This extensive exploration delves into the profound impact of edge computing in Industry 4.0, meticulously analysing its applications, advantages, challenges, and implications for the future of smart manufacturing and beyond.

8.2 Edge Computing: Pioneering the Next Era of Industry 4.0

At the vanguard of the Industry 4.0 revolution stands edge computing, poised to redefine the very essence of data processing and analytics by introducing a distributed computing paradigm that brings computational resources closer to the point of data generation. This transformative approach enables enterprises to mitigate latency, minimise bandwidth usage, and enhance privacy and security measures by deploying computing resources at the network's edge, whether within the physical confines of industrial machinery or at the periphery of the network infrastructure. These advancements represent a pivotal breakthrough for real-time data analysis and decision-making, essential for a myriad of critical applications ranging from predictive maintenance and quality control to process optimization and anomaly detection. By harnessing the power of edge computing, manufacturers can glean actionable insights from sensor data and telemetry streams in real-time, enabling proactive maintenance interventions, adaptive process adjustments, and rapid responses to emerging challenges on the factory floor. Moreover, the inherent capabilities of edge computing extend beyond mere real-time analytics to encompass offline operation and autonomy, ensuring uninterrupted operations and continuity of service even in environments with limited or intermittent connectivity.

This resilience to network disruptions is particularly critical in industrial settings where downtime can have significant financial repercussions and operational ramifications. Through the seamless integration of edge computing into the fabric of Industry 4.0, enterprises can unlock new levels of efficiency, agility, and innovation, driving sustainable growth and competitiveness in an increasingly digital and interconnected world. However, the widespread adoption of edge computing also presents formidable challenges and complexities that must be addressed to realise its full potential.

Chief among these is the need to ensure interoperability and compatibility between heterogeneous edge devices, sensors, and platforms, which often operate on disparate protocols and standards. Additionally, the decentralised nature of edge computing introduces new security risks and vulnerabilities, necessitating robust authentication, encryption, and access control mechanisms to safeguard sensitive data and assets from cyber threats and attacks.

Furthermore, the proliferation of edge computing architectures requires a rethinking of traditional approaches to data management, storage, and governance in industrial environments. As data is generated and processed at the edge, organisations must develop strategies for data aggregation, synchronisation, and consistency to ensure reliability and integrity across distributed edge nodes. This entails the adoption of scalable and resilient data management frameworks that can accommodate the unique requirements and constraints of edge computing deployments while adhering to regulatory and compliance mandates.

Despite these challenges, the transformative potential of edge computing in Industry 4.0 remains undeniable. By pioneering a distributed computing paradigm that brings intelligence closer to the point of action, edge computing empowers enterprises to unlock new levels of efficiency, agility, and innovation, driving sustainable growth and competitiveness in the digital age. However, realising this vision requires a concerted effort to address the technical, organisational, and regulatory challenges inherent in the adoption and integration of edge computing within the industrial ecosystem. Only through collaborative innovation, strategic foresight, and a steadfast commitment to security and reliability can we fully harness the transformative potential of edge computing to shape the future trajectory of manufacturing and industrial processes in the era of Industry 4.0.

8.3 Bringing Intelligence Closer: The Crucial Role of Edge Computing in Industry 4.0

The pivotal role of edge computing in the landscape of Industry 4.0 cannot be overstated, as it heralds a paradigm shift in the way data is processed, analysed, and acted upon within the industrial ecosystem. By bringing intelligence closer to the physical realm, edge computing revolutionises the speed and efficiency with which insights are generated, responses are formulated, and actions are executed. Through the localization of data processing at the edge of the network, enterprises can effectively slash latency and response times, facilitating real-time monitoring and control of manufacturing processes with unprecedented precision and immediacy. This capability proves particularly crucial in time-sensitive applications such as machine monitoring, where even minor delays in data processing could translate into substantial downtime, production losses, and operational inefficiencies. The inherent architecture of edge computing empowers localised decision-making, eliminating the need to transmit vast quantities of raw data to centralised cloud servers for processing and analysis.

Instead, edge devices and sensors autonomously analyse data streams in situ, leveraging predefined algorithms and machine learning models to derive actionable insights and initiate timely interventions without relying on external infrastructure. This not only amplifies operational efficiency and reliability but also reduces the strain on network bandwidth and minimises dependencies on stable internet connectivity, thereby enhancing resilience and adaptability in dynamic industrial environments. Moreover, by leveraging edge computing capabilities, enterprises can unlock new opportunities for innovation and optimization across the manufacturing value chain. From predictive maintenance and quality control to process optimization and supply chain management, edge computing empowers manufacturers to leverage real-time data analytics and predictive modelling to identify patterns, anticipate trends, and optimise operational performance with unparalleled agility and precision.

This transformative potential extends beyond the confines of individual factories and production lines, encompassing broader applications such as smart cities, autonomous vehicles, and decentralised energy grids, where edge computing plays a pivotal role in enabling localised intelligence, autonomy, and responsiveness. However, despite its myriad advantages, the widespread adoption of edge computing in the context of Industry 4.0 also poses significant challenges and complexities that must be addressed to realise its full potential. Chief among these is the need for interoperability and compatibility between heterogeneous edge devices, sensors, and platforms, which often operate on disparate protocols and standards. Additionally, the decentralised nature of edge computing introduces new security risks and vulnerabilities, necessitating robust authentication, encryption, and access control mechanisms to safeguard sensitive data and assets from cyber threats and attacks. Furthermore, the proliferation of edge computing architectures requires a rethinking of traditional approaches to data management, storage, and governance in industrial environments. As data is generated and processed at the edge, organisations must develop strategies for data aggregation, synchronisation, and consistency to ensure reliability and integrity across distributed edge nodes. This entails the adoption of scalable and resilient data management frameworks that can accommodate the unique requirements and constraints of edge computing deployments while adhering to regulatory and compliance mandates. Despite these challenges, the transformative potential of edge computing in Industry 4.0 remains undeniable. By pioneering a distributed computing paradigm that brings intelligence closer to the point of action, edge computing empowers enterprises to unlock new levels of efficiency, agility, and innovation, driving sustainable growth and competitiveness in the digital age. However, realising this vision requires a concerted effort to address the technical, organisational, and regulatory challenges inherent in the adoption and integration of edge computing within the industrial ecosystem. Only through collaborative innovation, strategic foresight, and a steadfast commitment to security and reliability can we fully harness the transformative potential of edge computing to shape the future trajectory of manufacturing and industrial processes in the era of Industry 4.0.

8.3.1 Edge of Innovation: Fostering Industry 4.0 Advancements through Edge Computing:

The advent of edge computing heralds a new era of innovation within Industry 4.0, empowering enterprises with novel applications and capabilities previously deemed impractical or unattainable. By deploying computing resources closer to the point of action, enterprises unlock opportunities for automation, autonomy, and intelligence. For instance, edge computing facilitates real-time analysis of sensor data to identify anomalies and trigger automated responses, such as adjusting equipment settings or scheduling maintenance activities. Similarly, edge-based AI models can scrutinise video feeds to detect defects or safety hazards in real-time, enhancing quality control and worker safety.

8.3.2 Decentralising Intelligence: Edge Computing's Influence on Industry 4.0:

Edge computing epitomises the decentralisation of intelligence in Industry 4.0, dispersing compute resources across the network to enable localised data processing and analysis. This paradigm shift reduces dependence on centralised cloud servers, fostering greater resilience, scalability, and flexibility. By disseminating intelligence to the edge, enterprises can diminish latency, amplify reliability, and ensure operational continuity, even in environments characterised by limited or sporadic connectivity. Moreover, edge computing empowers edge devices to operate autonomously, making real-time decisions sans reliance on cloud-based services or internet connectivity.

8.3.3 Real-Time Decision Making: The Crucial Role of Edge Computing in Industry 4.0 Systems:

Critical to Industry 4.0 systems, edge computing plays an instrumental role in facilitating real-time decision-making, allowing enterprises to analyse data and act on insights with minimal delay. By processing data locally at the edge, enterprises can truncate latency and response times, enabling swift insights and actions. This holds particular significance in time-sensitive applications such as predictive maintenance, where timely intervention can preclude costly equipment failures and production disruptions. Additionally, edge computing enables enterprises to winnow and prioritise data locally, reducing the volume of data transmitted to centralised cloud servers for processing, thereby refining efficiency and diminishing bandwidth usage.

8.3.4 Edge to Enterprise: Enmeshing Edge Computing into Industry 4.0 Strategies:

Efficiently integrating edge computing into Industry 4.0 strategies mandates a comprehensive approach that addresses technical, organisational, and cultural hurdles. From a technical standpoint, enterprises must deploy edge computing infrastructure that is scalable, secure, and interoperable with extant systems and processes. This may entail deploying edge gateways, edge servers, and edge analytics platforms to gather, process, and analyse data locally. Furthermore, enterprises must devise edge computing architectures that facilitate seamless integration with centralised cloud services, enabling hybrid deployments that harness the strengths of both edge and cloud computing.

8.3.5 Edge Intelligence: Revolutionising Manufacturing in Industry 4.0:

The advent of edge intelligence signifies a revolutionary shift in manufacturing within the context of Industry 4.0, heralding a new era characterised by real-time insights, predictions, and actions at the very point of data generation. Through the deployment of advanced artificial intelligence (AI) models and algorithms at the edge of the network, enterprises can analyse data locally, enabling autonomous decision-making devoid of constant reliance on centralised cloud services. This decentralised approach not only reduces latency and network congestion but also fosters swiffer responses to changing conditions, heightening operational efficiency and agility. By scrutinising sensor data in real-time, edge-based AI models can identify anomalies, predict equipment failures, and optimise production schedules, facilitating proactive maintenance and resource allocation. Despite challenges such as interoperability and security concerns, the transformative potential of edge intelligence in Industry 4.0 is undeniable, offering manufacturers unprecedented levels of autonomy, efficiency, and competitiveness in an increasingly digital and interconnected world.

8.4 Empowering Smart Factories: The Catalytic Role of Edge Computing in Industry 4.0 Success

At the crux of Industry 4.0 lies the indispensable role of edge computing, serving as a linchpin for success by endowing smart factories with the capability to harness real-time insights, automation, and autonomy. Through the strategic deployment of computing resources at the network's edge, enterprises can engage in meticulous scrutiny of data locally, affording them the ability to make intelligent decisions in the blink of an eye, all without the cumbersome reliance on centralised cloud services.

This paradigm shifts not only slashes latency and reduces network congestion but also engenders a newfound agility and responsiveness, imperative qualities for navigating the ever-evolving terrain of today's fast-paced manufacturing landscape. By capitalising on edge computing's prowess, smart factories can achieve augmented efficiency and productivity, as they seamlessly integrate real-time data analytics into their operational fabric, enabling rapid adjustments and optimizations to processes, workflows, and resource allocations. The transformative capabilities of edge computing extend beyond mere efficiency gains to encompass the realm of autonomy, as smart factories emerge empowered to operate autonomously, making informed decisions and adjustments grounded in local conditions and exigencies.

This newfound autonomy curtails the reliance on human intervention, enabling uninterrupted operation even in the face of unforeseen disruptions or challenges. Through the convergence of real-time insights, automation, and autonomy facilitated by edge computing, smart factories are poised to chart a course towards unprecedented levels of efficiency, resilience, and competitiveness, as they navigate the complexities of Industry 4.0 with confidence and dexterity. However, the widespread adoption of edge computing also presents formidable challenges and complexities that must be navigated to realise its full potential. Chief among these is the imperative to ensure interoperability and compatibility between heterogeneous edge devices, sensors, and platforms, which often operate on disparate protocols and standards. Additionally, the decentralised nature of edge computing introduces new security risks and vulnerabilities, necessitating robust authentication, encryption, and access control mechanisms to safeguard sensitive data and assets from cyber threats and attacks. Furthermore, the proliferation of edge computing architectures necessitates a rethinking of traditional approaches to data management, storage, and governance in industrial environments. As data is generated and processed at the edge, organisations must develop strategies for data aggregation, synchronisation, and consistency to ensure reliability and integrity across distributed edge nodes. This entails the adoption of scalable and resilient data management frameworks that can accommodate the unique requirements and constraints of edge computing deployments while adhering to regulatory and compliance mandates. Despite these challenges, the transformative potential of edge computing in Industry 4.0 remains undeniable. By pioneering a distributed computing paradigm that brings intelligence closer to the point of action, edge computing empowers enterprises to unlock new levels of efficiency, agility, and innovation, driving sustainable growth and competitiveness in the digital age. However, realising this vision requires a concerted effort to address the technical, organisational, and regulatory challenges inherent in the adoption and integration of edge computing within the industrial ecosystem. Only through collaborative innovation, strategic foresight, and a steadfast commitment to security and reliability can we fully harness the transformative potential of edge computing to shape the future trajectory of manufacturing and industrial processes in the era of Industry 4.0

8.5 Reshaping the Industrial Landscape: Edge Computing's Contribution to Industry 4.0

As a key component of Industry 4.0, edge computing is altering the industrial environment and acting as a catalyst for the introduction of new services, value propositions, and business models that rethink the fundamentals of organisational operations. By drawing intelligence closer to the point of action, edge computing presents enterprises with an array of fresh opportunities for innovation and differentiation. One of the primary implications of edge computing lies in its ability to empower enterprises to provide innovative services that cater to the evolving needs of the market. For instance, by leveraging edge computing capabilities, enterprises can offer predictive maintenance services, remote monitoring solutions, and real-time analytics platforms to their clients. These solutions empower clients to optimise their operations, enhance reliability, and reduce costs by leveraging real-time insights derived from data collected at the edge. Through the integration of edge computing technologies, enterprises can analyse sensor data locally, enabling them to make intelligent decisions in real-time without relying on centralised cloud services. This localization of data processing not only reduces latency and network congestion but also fosters agility and responsiveness, crucial attributes for thriving in today's fast-paced manufacturing landscape. Moreover, edge computing enables enterprises to monetize their data assets by offering data-as-a-service solutions to clients. These solutions allow clients to access and analyse industrial data in real-time, unlocking new revenue streams and business opportunities for enterprises. By capitalising on edge computing infrastructure, enterprises can cater to the growing demand for actionable insights and intelligence, thereby enhancing their competitiveness in the digital age. Furthermore, the deployment of edge computing facilitates the development of tailored solutions and value-added services that address the specific needs of different industries and verticals. Whether in manufacturing, logistics, healthcare, or energy, edge computing empowers enterprises to customise their offerings to meet the precise requirements of their clients, fostering deeper engagement and satisfaction. For instance, in the manufacturing sector, edge computing enables enterprises to deploy customised solutions for process optimization, quality control, and supply chain management, gaining a competitive edge in the market. Similarly, in the healthcare industry, edge computing facilitates the development of personalised healthcare solutions, remote patient monitoring platforms, and predictive analytics tools, improving patient outcomes and experiences. However, despite the numerous benefits and opportunities offered by edge computing, its widespread adoption also poses challenges and complexities that must be addressed to realise its full potential.

These challenges include ensuring interoperability and compatibility between heterogeneous edge devices, sensors, and platforms, as well as addressing security risks and vulnerabilities associated with decentralised data processing. Additionally, organisations must develop strategies for data management, storage, and governance to ensure reliability and integrity across distributed edge nodes. Despite these challenges, the transformative potential of edge computing in Industry 4.0 remains undeniable. By empowering enterprises to unlock new levels of efficiency, agility, and innovation, edge computing is poised to shape the future of the industrial landscape, driving sustainable growth and competitiveness in the digital era.

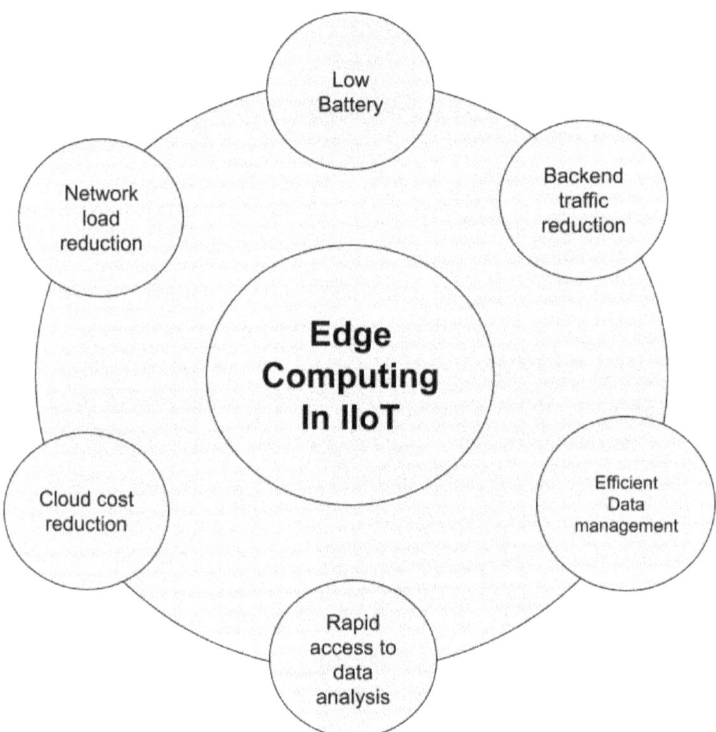

Exhibit 8.2 Benefits of Edge Computing

Edge-Centric Future: Redefining Industry 4.0 through Edge Computing

In the unfolding narrative of Industry 4.0, the trajectory is unmistakably imbued with an edge-centric ethos, positioning edge computing as a linchpin in enabling intelligent, autonomous, and interconnected manufacturing processes. With its capability to usher intelligence closer to the point of action, edge computing emerges as a critical facilitator for swift insights, responses, and actions, essential for navigating the complexities of today's expeditious business milieu. This shift towards edge-centricity not only accelerates decision-making but also fosters agility and adaptability, prerequisites for success in the dynamic landscape of modern industry. Moreover, edge computing empowers enterprises to seize novel opportunities for innovation and differentiation, fuelling continual improvement and value creation across the manufacturing value chain. By leveraging edge computing capabilities, organisations can unlock new avenues for optimising processes, enhancing productivity, and delivering superior products and services to customers. As the progress towards the Fourth Industrial Revolution, the centrality of edge computing is set to burgeon further, reshaping the trajectory of manufacturing and industrial processes. This comprehensive overview underscores the pivotal role of edge computing in Industry 4.0, encapsulating its diverse applications, myriad benefits, formidable challenges, and far-reaching implications for the future of smart manufacturing and beyond. By harnessing the full potential of edge computing, enterprises can unlock fresh opportunities for innovation, efficiency, and competitiveness in the Fourth Industrial Revolution, positioning themselves for sustained success and growth in the digital age.

Key Notes:

Topic	Keynotes
Significance of Edge Computing	Edge computing revolutionizes industrial processes by enabling real-time decision-making.
	It brings intelligence closer to the point of action, driving efficiency and innovation in manufacturing.
Pioneering the Next Era of Industry 4.0	Edge computing mitigates latency and enhances security by deploying computational resources at the network's edge.
	Real-time analytics empower proactive maintenance and adaptive process adjustments.
Bringing Intelligence Closer	Edge computing reduces latency, enabling swift insights and actions in time-sensitive applications.
	It fosters autonomy and real-time decision-making without reliance on cloud services.

Topic	Keynotes
Edge of Innovation	Edge computing facilitates automation, autonomy, and real-time analysis of sensor data.
	It enables real-time insights and predictions, enhancing operational efficiency and agility.
Decentralising Intelligence	Edge computing disperses compute resources across the network, reducing dependence on centralised servers.
	It empowers edge devices to operate autonomously, making real-time decisions without reliance on cloud services.
Real-Time Decision Making	Edge computing facilitates real-time decision-making by processing data locally, minimizing latency.
	It enables enterprises to prioritize and act on insights swiftly, crucial in time-sensitive applications.
Edge to Enterprise	Enterprises must address technical, organizational, and cultural hurdles to integrate edge computing effectively.
	Edge computing architectures should seamlessly integrate with centralised cloud services for hybrid deployments.
Edge Intelligence	Edge intelligence enables real-time insights and predictions, enhancing operational efficiency and agility.
	It allows enterprises to analyse data locally, reducing reliance on centralized cloud services.
Reshaping the Industrial Landscape	Edge computing introduces new services and business models, leveraging real-time insights for value creation.
	Enterprises can customize solutions across industries, fostering deeper engagement and satisfaction.

Summary

The chapter delves into the transformative role of edge computing within the framework of Industry 4.0, illuminating how it reshapes manufacturing and industrial processes with unprecedented efficiency and agility. By bringing intelligence closer to the point of action, edge computing enables real-time decision-making, proactive maintenance, and adaptive process adjustments. Despite challenges like interoperability and security risks, the chapter underscores the undeniable transformative potential of edge computing in driving innovation and competitiveness in the digital era, heralding a new era of autonomy, resilience, and competitiveness in the digital age.

Trivia Challenge: Industry 4.0

Welcome to the Industry 4.0 Trivia Challenge! Get ready to test your knowledge of the wacky world of Industry 4.0 with these hilarious multiple-choice questions. Each question is paired with some side-splitting facts and figures about Industry 4.0. Let us dive in and have a giggle!

1) What is Edge Computing's role in Industry 4.0 like in a superhero movie?

> A) A sidekick in a cape, trailing behind Industry 4.0's hero.

> B) The unsung hero pioneering the next era of Industry 4.0!

> C) The villain trying to decentralize intelligence for chaos.

> D) The mysterious figure lurking at the edge of industrial landscapes.

2) How does Edge Computing bring intelligence closer to the source in Industry 4.0?

> A) By installing brain implants in factory workers.

> B) By pioneering real-time decision-making at the edge.

> C) By teleporting factories to the edge of the universe.

> D) By whispering secrets of innovation to the machinery.

3) What does Edge Computing do to decentralize intelligence within Industry 4.0 systems?

> A) Transforms factories into independent thinkers.

> B) Spreads intelligence across the industrial landscape.

> C) Creates a network of rebellious factory robots.

> D) Turns factories into puzzle pieces for a giant industrial jigsaw.

4) How does Edge Computing empower smart factories according to the chapter?

> A) By giving them superhero capes for added efficiency.

> B) By enabling real-time decision-making and innovation.

> C) By hosting edge-of-the-universe factory parties.

> D) By teaching machinery, the art of wit and charm.

5) **According to the chapter, what adjective describes Edge Computing's role in Industry 4.0?**

   A) Edge-tastic.

   B) Pivotal.

   C) Mystical.

   D) Revolutionary.

6) **What does Edge Computing do to reshape the industrial landscape?**

   A) Carves it into a giant edge-shaped sculpture.

   B) Revolutionizes manufacturing with Edge Intelligence.

   C) Plants edge-shaped trees along factory pathways.

   D) Gives factories a fashionable edge in the competitive market.

7) **How does Edge Computing revolutionize manufacturing, as described in the chapter?**

   A) By teaching factory workers to dance at the edge of innovation.

   B) By enabling real-time decision-making and empowering smart factories.

   C) By turning factories into edge-of-the-universe amusement parks.

   D) By adding a touch of mystery and intrigue to industrial processes.

8) **What metaphor does the chapter use to describe Edge Computing's role in Industry 4.0?**

   A) The unsung hero pioneering the next era of Industry 4.0.

   B) The mysterious figure lurking at the edge of industrial landscapes.

   C) The edge-shaped puzzle piece fitting perfectly into the Industry 4.0 picture.

   D) The rebel spreading intelligence like wildfire across factories.

9) **According to the chapter, how does Edge Computing bring innovation to Industry 4.0?**

   A) By hosting edge-themed costume parties for factory workers.

   B) By enabling real-time decision-making and fostering innovation.

   C) By painting edge-shaped murals on factory walls.

   D) By teaching machinery, the art of creativity and imagination.

10) What adjective does the chapter describe Edge Computing's role in enabling real-time decision-making?

A) Edge-tastic.

B) Crucial.

C) Mysterious.

D) Enlightening.

Abbreviation	Terms
IoT	Internet of Things
AI	Artificial Intelligence
CPS	Cyber-Physical Systems
ECP	Edge Computing Platform
ML	Machine Learning
QC	Quality Control
PM	Predictive Maintenance

Chapter 9

Big Data Analytics and Software-Defined Networks: Industry 4.0

Learning Objectives:

- Understand the pivotal role of big data analytics and software-defined networks (SDN) in Industry 4.0 transformations.

- Comprehend the significance of data analytics in industrial applications, particularly in the context of the Industrial Internet of Things (IIoT).

- Explore the fundamentals of software-defined networking (SDN) and its impact on network flexibility in industrial settings.

- Gain insights into predictive maintenance techniques and their applications in enhancing manufacturing efficiency and reducing downtime.

- Learn about supply chain optimization strategies enabled by data analytics and machine learning algorithms.

9.1 Role of Data Analytics in Industry 4.0

The Industrial Internet of Things (IIoT) is an extension of the Internet of Things (IoT) for industrial and manufacturing applications. It entails connecting smart devices, sensors, and other industrial equipment to a network in order to gather, share, and analyse data. Unlike consumer-oriented IoT, which focuses on improving everyday living through smart home gadgets or wearables, IIoT is intended to optimise and simplify industrial operations. IIoT promotes the development of smart factories and industrial ecosystems in which machines, systems, and humans can all connect and work seamlessly. Sensors, actuators, networking solutions, and sophisticated analytics tools are essential components of the IIoT. These components operate together to collect real-time data from several sources in an industrial context, allowing for data-driven decision-making and automation.

Step	Description
1. Set a Strategy	Define the objectives and goals of the big data analytics initiative. Identify the relevant sources of big data to address these objectives.

Step	Description
2. Identify Big Data Sources	Determine the sources of big data, which could include sensors, social media, customer databases, logs, etc.
3. Access, Manage and Store Big Data	Establish processes and infrastructure to access, manage, and store large volumes of data efficiently. Utilize technologies like Hadoop, NoSQL databases, or cloud storage solutions.
4. Analyse	Apply analytical techniques such as statistical analysis, machine learning, or data mining to extract valuable insights and patterns from the data.
5. Make Intelligent Decisions	Interpret the analytical findings and use them to make informed decisions, optimise processes, or create new strategies.

Exhibit 9.1 Big Data Analysis Process Steps

9.1.1 Significance of IIoT in the Industrial Sector:

Operational Efficiency: IIoT plays a crucial role in enhancing operational efficiency by providing real-time insights into the performance of machinery and processes. This allows for predictive maintenance, reducing downtime and optimising resource utilisation.

Cost Reduction: Through predictive maintenance and improved efficiency, IIoT helps in reducing operational costs. Organisations can identify and address issues before they escalate, preventing costly breakdowns and unnecessary maintenance.

Data-Driven Decision Making: The IIoT creates massive volumes of data that may be analysed to make sound judgements. This data-driven approach enables better strategic planning, resource allocation, and overall optimization of industrial processes. Supply Chain

Optimization: IIoT facilitates end-to-end visibility in the supply chain. This includes tracking the movement of raw materials, monitoring inventory levels, and optimising production schedules, leading to a more agile and responsive supply chain.

Quality Control: Real-time monitoring of production processes through IIoT ensures high-quality output by identifying and addressing deviations from quality standards promptly.

Safety and Compliance: IIoT helps in ensuring a safer working environment by monitoring equipment conditions and detecting potential hazards. It also aids in compliance with industry regulations through accurate data recording and reporting.

Industrial IoT (IIoT) delivers operational efficiency gains by providing real-time performance insights into machinery and processes, enabling predictive maintenance approaches that reduce downtime and optimise resource utilisation. Through predictive maintenance and improved efficiency, IIoT also helps reduce operational costs by identifying and addressing issues before they escalate and cause costly breakdowns or unnecessary maintenance.

IIoT generates vast amounts of machine data that can be analysed to support data-driven decision making, facilitating better strategic planning, resource allocation, and optimization of industrial processes. It also facilitates end-to-end supply chain visibility through monitoring activities like raw material movement, inventory levels, and production scheduling, leading to a more agile and responsive supply chain overall. Real-time process monitoring with IIoT further ensures high-quality output by identifying and addressing any deviations from quality standards in a timely manner. Finally, IIoT helps ensure workplace safety by monitoring equipment conditions and detecting potential hazards, while also aiding regulatory compliance through accurate data recording and reporting activities.

9.2 Network Flexibility Software-defined networking (SDN)

It provides a flexible and programmable network infrastructure, enabling dynamic allocation of resources based on changing demands from industrial internet of things (IIoT) applications. Security SDN enhances security by facilitating centralised control and monitoring of network traffic. This is critical in IIoT environments where security is a top priority to protect sensitive industrial data and ensure integrity of operations. Scalability SDN supports scalability requirements of IIoT by allowing efficient management and scaling of network resources as the number of connected devices and data traffic increases. The integration of IIoT, big data analytics, and SDN has transformative effects on the industrial sector, leading to increased efficiency, improved decision-making, and a more responsive and secure industrial ecosystem.

The Industrial Internet of Things (IIoT) generates massive amounts of data from connected industrial devices, and analytics plays a vital role in extracting valuable insights from this data. IIoT analytics help organisations transform raw data into actionable intelligence, allowing them to make informed decisions, improve operational efficiency, and achieve better overall outcomes. With the help of analytics, businesses can identify patterns, trends, and anomalies in real-time and historical data, leading to predictive and prescriptive insights that can help improve performance, reduce downtime, and optimise resource utilisation.

Real-time data processing in IIoT (Industrial Internet of Things) analytics refers to the immediate analysis of data as it is generated. This enables organisations to respond rapidly to changing conditions. Real-time analytics is of great importance because it allows for quick decision-making, particularly in situations where timely responses are crucial. It is essential for monitoring and controlling industrial processes in real-time, identifying anomalies, and taking immediate corrective actions.

Predictive analytics in Industrial Internet of Things (IIoT) refers to the use of historical and real-time data to predict future events or conditions. This technique helps in anticipating equipment failures, maintenance needs, and other potential issues before they occur.

The importance of predictive analytics lies in its ability to facilitate proactive maintenance strategies, thereby reducing downtime and extending the lifespan of equipment. It also aids in optimising resource allocation, improving efficiency, and ultimately lowering operational costs.

Prescriptive analytics is a data analysis approach that not only predicts future outcomes but also recommends actions to optimise processes and achieve specific goals. This methodology combines predictive insights with decision optimization to suggest the best course of action.

Prescriptive analytics is essential as it guides decision-makers in selecting the most effective actions to achieve their desired outcomes. It is particularly valuable in complex industrial environments where multiple variables need to be considered for optimal decision-making.

Step	Description
Gather Data	Collect data from Programmable Logic Controllers (PLC), sensors, and other devices/sources.
Store & Clean Data	Determine the data attributes and store the data in a structured manner. Cleanse and pre-process the data to ensure accuracy and consistency.
Analyse Data	Identify suitable analytical models and algorithms. Apply these techniques to analyse the data, extract intelligence, and make predictions or insights.

Step	Description
Display Data	Present the analysed data in various formats such as a Machine Health Index, Web Dashboard, or Mobile application for easy interpretation and decision-making.

Exhibit 9.2 Predictive Maintenance

9.3 Information Pre-processing and Highlight Engineering

Information Pre-processing:

Definition: Information pre-processing includes cleaning and changing crude information into an arranged reasonably for examination. This incorporates taking care of lost values, expelling exceptions, and normalising information to guarantee consistency and reliability.

Significance: In IIoT, information pre-processing is pivotal to guarantee the precision and unwavering quality of models. It makes strides in the quality of input information, driving to more strong and exact insights.

9.4 Building and Preparing Models for IIoT Applications

Relapse Models:

Application: In IIoT, relapse models are utilised for errands like foreseeing gear disappointment times or evaluating the remaining valuable life of the apparatus. They analyse verifiable information to recognize designs and trends.

Case: Straight relapse can be utilised to anticipate the time until the following upkeep is based on components such as temperature, utilisation, and verifiable support data.

Classification Models:

Application: Classification models in IIoT are utilised for tasks such as blame discovery or quality control. These models classify information into predefined categories based on learned patterns.

Example: A classification demonstration seems to distinguish whether an item on a generation line meets quality standards or classify sensor readings as ordinary or characteristic of a potential issue.

Clustering Algorithms:

Application: Clustering calculations are utilised in IIoT for errands like inconsistency location or gathering comparable resources or forms together.

Case: K-means clustering can be utilised to bunch comparative machines based on their execution characteristics, supporting focused on upkeep techniques for particular machine groups.

The part of information science in IIoT is essential for turning crude information into significant bits of knowledge. Information pre-processing guarantees information quality including building upgrades, the significance of factors, and the application of relapse, classification, and clustering models to address particular challenges within the industrial landscape. The proceeded advancement of information science and machine learning strategies in IIoT holds the guarantee of advanced optimising mechanical processes and driving advancement within the fabricating and generation sectors.

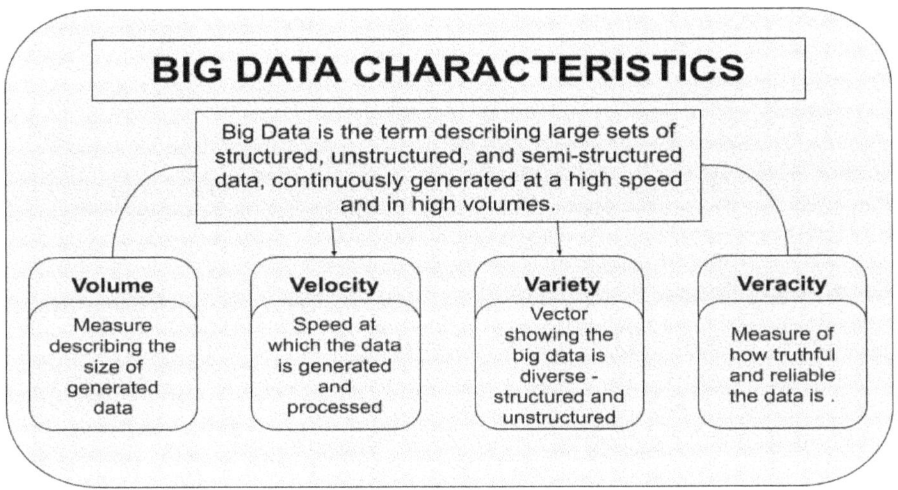

Exhibit 9.3 Big Data Structure

9.5 Artificial Intelligence Overview

Definition: Artificial Intelligence refers to the creation of computer systems that can perform tasks that often require human intelligence. These activities include learning, reasoning, problem solving, comprehension, and language comprehension.

Basic Principle: Artificial intelligence works on principles such as machine learning, where machines learn from data and systems are intelligently designed to improve performance.

Machine Learning & Deep Learning

Machine Learning (ML): ML is a category of artificial intelligence that enables machines to learn from data without being programmed. It contains an algorithm that improves its performance over time.

Deep Learning: Deep learning is a special type of machine learning inspired by the structure and function of the human brain's neural network. It specialises in managing information and standards.

Cross-sector applications of AI

Medical diagnosis

Artificial Intelligence algorithms analyse medical images and help diagnose diseases such as cancer through information about the images.

Diagnostic Chabot's use natural language processing (NLP) to interact with patients and provide initial assessments.

Self-Driving

Artificial Intelligence strengthens the decision-making process of driverless cars, allowing them to navigate, detect obstacles and make driving decisions.

Machine learning algorithms improve the car's ability to adapt to road changes and improve the road.

Natural Language Processing in Customer Service

AI-powered chatbots use NLP to understand and respond to customer questions and provide effective and personalised support.

AI-powered virtual assistants can hold conversations to help users complete various tasks.

Role in Business Development

Automation of Repetitive Tasks

Artificial Intelligence automates routine and repetitive tasks, freeing up time for human resources to engage in more complex and creative tasks.

Robotic Process Automation (RPA) uses artificial intelligence to mimic human behaviour in the use of software to make tasks more efficient.

Predictive Analytics for Decision Making

Artificial Intelligence algorithms analyse big data to identify patterns and trends, allowing organisations to make data-driven decisions.

Analytical models predict future outcomes and assist in strategic planning and risk management.

Fair decision making in Artificial Intelligence Bias and Fairness

Artificial Intelligence systems will be subject to bias in training data, resulting in bias. Ensuring fairness in AI applications requires careful evaluation and reducing bias.

Ethical AI development aims to eliminate injustice and promote equal representation.

Transparency and Accountability

The lack of transparency in artificial intelligence decision-making processes is a cause for concern. We strive to build transparent AI models to provide insight into decision-making processes.

Establish a system of accountability to account for the use of skills and provide clear guidelines on responsibility for behaviour and errors.

The future development of artificial intelligence

descriptive intelligence, focuses on making it easier to understand and explain the decision-making process of intelligence.

Greater transparency increases user trust and increases compliance.

Human and Artificial Intelligence Collaboration

The future development of artificial intelligence demonstrates the cooperation between humans and artificial intelligence by utilising single muscle power. The human-machine intelligence cycle includes human care to ensure the fair and responsible use of artificial intelligence.

As we explore the evolving landscape of artificial intelligence, it is clear that its impact extends beyond technological advances. Ethical considerations in AI development highlight the need to take responsibility to ensure AI technology contributes to society. The future promises a more transparent, collaborative and descriptive intelligence that enhances human potential while maintaining ethical standards.

9.6 Big data

Definition and characteristics

Quantity, speed, diversity

Quantity: Big data refers to the amount of information produced on an unprecedented scale. It involves processing and analysing data sets that exceed the capabilities of traditional systems.

Speed: Big data is characterised by the speed at which data is created, written and processed. Instant or near-volume analysis is often required.

Diversity: Big data includes many types of data, including structured, semi-structured and unstructured data. This difference creates problems in storage and identification.

Applications in Business and other fields

Analytical Research

Big data analyses management patterns and models to make predictions from historical data. Organisations use forecasts to predict future outcomes and make operational decisions.

Industries such as finance, business and healthcare use predictive analytics to optimise strategies and reduce risk.

Customer Relationship Management

The role of big data plays an important role in Customer Relationship Management (CRM). Analysing customer interactions and behaviour helps businesses understand preferences, improve customer experience and provide personalised services.

CRM platforms use big data to create a 360-degree view of customers, enable better communication and build relationships.

Challenges of Big Data Management

Data Security and Privacy

The volume and diversity of big data make it a target of security breaches. Data security and privacy are quite complex.

Complying with data protection laws such as GDPR requires effective management of the security and integrity of data.

Data Quality and Integration

Managing data quality is a difficult task in a big data environment with a lot of data and models.

Integrating disparate data creates challenges and requires tools and processes to ensure consistent and accurate data.

Linking disparate data poses challenges that require tools and processes to ensure consistent and accurate data.

Future Trends in Big Data

Edge Computing and Instant Processing

Edge computing brings processing power closer to the data source, reducing latency and enabling instant analysis. This model is important for applications that require rapid decision making.

Rapid processing of big data accelerates change by enabling organisations to gain insight and respond quickly to changes.

Technology Advanced Data Analysis Technology

Advanced Data Analysis Technology

Machine learning and artificial intelligence enable complex data analysis technology.

Technologies such as natural language processing, sentiment analysis, and graph analysis can increase the depth and breadth of big data insights.

As big data grows and we delve deeper into the field, it is clear that its applications extend far beyond data storage and retrieval. Big data is a new power that enables organisations to make informed decisions, improve user experience and gain competitive advantage. However, issues related to data security, efficiency and privacy highlight the importance of ethics and responsibility in the management of big data. The future will not only increase the value of knowledge, but will also bring the development of technology and ideas that will eliminate the need for a large and powerful field.

Advanced Analysis

Data Exploration

Definition: Data exploration involves discovering patterns and insights in large data sets using a variety of techniques such as analytics, machine learning, and artificial intelligence.

Purpose: Data mining helps discover relationships and patterns in data and supports informed decision making.

Text Analysis

Definition: Text analysis, also known as text mining, involves extracting themes and patterns from unrelated data. This includes factor analysis, factor analysis and modelling.

Application: Text analytics is used to analyse customer feedback, social media content and data collected from various industries.

Business Intelligence and Decision Making

Dashboards and Visualization Tools

Dashboards: Dashboards provide graphical representations of key performance indicators (KPIs) and metrics. They offer users a simple and intuitive way to track and analyse data.

Visualisation tools: Advanced visualisation tools like Tableau and Power BI can help users create tables, charts, and dashboards to improve data understanding.

Prescriptive Analysis

Definition: Prescriptive analysis is the process of optimising results based on predictive models and historical data. Not only can it predict future events, but it can also suggest the best course of action.

Benefits: Helps organisations make better decisions by providing analytical writing, advice and recommendations.

Combining Machine Learning and Artificial Intelligence

Best Practices

Synergy with Predictive Analytics: Advanced analytics combined with predictive analytics to improve prediction towers and forecast models.

Machine Learning Integration: Machine learning algorithms help improve prediction accuracy and predictive ability by improving instant understanding of the model.

Continuous improvement and adaptation

Iterative Learning: Using machine learning for advanced analytics helps continuously learn from new data. This iterative process creates patterns and predictions.

Adaptive Decision Making: Organisations can adjust their strategies based on instantaneous understanding and respond dynamically to changes.

Supply Chain Optimization

Demand Forecasting: Advanced analysis helps predict demand patterns, improve product quality and improve product quality.

Risk Management: Identify and mitigate risks such as disruptions or delays in delivery through forecasting and analysis.

Fraud Detection and Prevention

Fraud Detection: Advanced analytics finds unusual patterns that lead to fraud in the financial sector.

Behavioural Analysis: Analyse behaviour patterns to identify situations and fraud in businesses such as finance and online platforms.

As we explore the world of advanced analytics, it is clear that its applications extend beyond traditional data analysis. The combination of data mining, text analysis, and visualisation tools gives organisations the tools to discover insights from their data. Integration with artificial intelligence and machine learning not only increases operational efficiency but also enables change and continuous improvement. Strategic use in supply chain optimization and fraud demonstrates the evolution of advanced analytics in informing decision-making processes across businesses.

9.7 Introduction to R and Julia Programming Languages

R Programming Language

R is a statistical programming language and software environment designed for statistical computing, data analysis, and graphics. It provides a wide array of statistical and graphical techniques and is widely used by statisticians, data scientists, and researchers. Features: R is known for its extensive collection of packages and libraries for statistical analysis, data visualisation, and machine learning. It is an open-source language with a large and active community contributing to its development.

Julia Programming Language

Julia is a high-level, high-performance programming language specifically designed for technical and scientific computing. It aims to provide a fast and efficient solution for numerical and data-intensive tasks while maintaining a user-friendly syntax.

Julia is known for its just-in-time (JIT) compilation, making it suitable for performance-critical applications. It is open-source, and its syntax is designed to be familiar and accessible to users of other technical computing environments.

Advantages of R and Julia in IIoT Analytics

R in IIoT Analytics: Rich Statistical Libraries: R excels in statistical analysis and provides a vast collection of libraries for conducting advanced analytics. This makes it well-suited for tasks such as predictive maintenance, quality control, and anomaly detection in IIoT.

Data Visualization: R offers powerful data visualisation capabilities, making it easier for analysts and data scientists to create informative and visually appealing charts and graphs based on IIoT data.

Julia in IIoT Analytics: Performance Optimization: Julia's JIT compilation allows for high-performance computing, making it suitable for computationally intensive tasks in IIoT analytics, such as processing large datasets and running complex machine learning algorithms. Ease of Use: Julia is designed to be user-friendly, with a syntax that is intuitive and similar to other technical computing languages. This makes it accessible for data scientists and engineers working on IIoT projects.

Case Studies Highlighting the Application of R and Julia in Industrial IoT Projects

R for Predictive Maintenance:

Overview: A manufacturing company leveraged R's statistical modelling capabilities for predictive maintenance across its production line. Historical data fed statistical models from R's extensive library to anticipate equipment failures.

Outcome: Implementation reduced unplanned downtime significantly. Machinery efficiency and optimised maintenance schedules also increased.

Julia for Real-Time Data Processing:

Overview: A utility company implemented Julia for real-time data processing within its smart grid system. Julia's performance optimizations allowed efficient processing and analysis of streaming data from smart metres.

Outcome: Real-time data supported faster decision-making, improved grid reliability, and enhanced energy distribution efficiency.

Integrated R and Julia for Supply Chain Optimization:

Overview: A logistics company integrated R and Julia in a hybrid analytics solution for supply chain optimization. R supported data analysis and visualisation while Julia handled computationally intensive optimization tasks.

Outcome: Integration resulted in a more responsive, optimised supply chain through improved route planning, inventory management, and overall cost reductions.

Both R and Julia languages serve important roles in Industrial IoT analytics. R specialises in statistical analysis and data visualisation. Julia's focus on high-performance computing makes it well-suited for computationally heavy tasks.

Case studies demonstrate successful applications across various Industrial IoT projects, showcasing versatility and impact within the industrial sector.

9.8 Overview of Hadoop in the Context of IIoT

Hadoop is an open-source framework designed to efficiently store and process large datasets in a distributed and scalable manner. In the context of Industrial Internet of Things (IIoT) applications, where vast amounts of data are generated from sensors, devices, and industrial processes, Hadoop provides a robust solution for data management and analytics needs. It enables organisations to effectively handle the volume, velocity, and variety of data generated by IIoT devices and systems.

Managing Large-Scale Data in IIoT Environments:

Scalability: Hadoop is designed to seamlessly expand its storage and processing capacity horizontally as the volume of IIoT data grows. This scalability is essential for addressing the massive datasets generated in industrial environments.

Fault Tolerance: Hadoop ensures fault tolerance by replicating data across multiple nodes in its distributed file system. This feature is crucial for IIoT applications where data reliability and availability are key priorities.

Batch Processing: Hadoop supports batch processing, making it suitable for scenarios in IIoT where large volumes of historical data need to be analysed for tasks such as predictive maintenance, quality control, and performance monitoring.

Industrial Internet of Things (IIoT) generates a massive amount of data that needs to be stored and managed reliably. Hadoop Distributed File System (HDFS) is used for this purpose. It allows the storage and retrieval of various types of industrial data such as sensor data and logs.

MapReduce is a programming model and processing engine used for distributed data processing in Hadoop. It involves two phases - the Map phase for data transformation and the Reduce phase for summarization. In IIoT, MapReduce is useful for batch processing tasks. For example, it can analyse historical sensor data to identify patterns, trends, or anomalies that can inform decision-making and optimise industrial processes.

Apache Spark is an open-source, fast, and general-purpose cluster computing system that supports more types of computations, including real-time processing and machine learning. In the Hadoop ecosystem, Apache Spark is used for scenarios where real-time processing is critical. It enables organisations to process and analyse streaming data from IIoT devices in real-time, allowing for timely insights and quicker decision-making.

Overall, Hadoop's distributed and scalable architecture is an excellent tool for managing and processing large-scale data in the context of IIoT. Its components, including HDFS, MapReduce, and Apache Spark, provide a comprehensive solution for storing, processing, and analysing industrial data. This contributes to enhanced decision-making and operational efficiency in industrial settings.

Ensuring Data Integrity and Confidentiality in IIoT Analytics

Issue: IIoT systems generate sensitive operational and equipment data. Maintaining data integrity and confidentiality is imperative to prevent disruptions, safety hazards, or intellectual property theft from unauthorised access or tampering.

Resolution: Implementing robust access controls, encryption, and secure transmission protocols addresses these concerns. Regular audits and monitoring detect and respond to potential security breaches.

Addressing Cybersecurity Threats

Issue: IIoT systems face risks from malware, ransomware, and phishing attacks. A successful cyberattack could severely impact data integrity and the safety and reliability of industrial processes.

Resolution: Employing best practices like regular software updates, intrusion detection, and network segmentation is essential. Organisations should also invest in personnel training and incident response planning.

Mitigating Device Vulnerabilities

Issue: IIoT devices often have limited resources, making them vulnerable to breaches. Devices could enable attackers to compromise the entire IIoT ecosystem.

Resolution: Implementing device-level security like firmware updates, secure boot, and embedded protocols helps mitigate vulnerabilities. Regular assessments and patch management are also critical.

Scalability Considerations for Industrial IoT Systems

Data Volume and Variety:

As the number of connected Industrial IoT devices increases, handling and processing the diverse data formats and sources generated becomes more complex. Implementing scalable storage solutions such as distributed file systems like Hadoop Distributed File System (HDFS), and utilising data processing frameworks like Apache Spark, can help efficiently manage large volumes of data.

Network Performance:

Industrial IoT systems rely on network connectivity to transmit data between devices and backend systems. The increasing number of connected devices can lead to network congestion and latency issues. Employing edge computing solutions to process data closer to the source, optimising network architecture, and utilising protocols designed for low-latency communication can address network performance challenges.

Computing Resources:

Industrial IoT analytics often require significant processing power, especially for real-time data analysis and machine learning tasks. Traditional computing architectures may struggle to keep up with the demands of processing large datasets in real time. Leveraging cloud computing resources, adopting distributed computing frameworks, and exploring specialised hardware for accelerative computing (e.g., GPUs) can enhance computing capabilities and scalability.

Key Notes:

Topic	Keynotes
Role of Data Analytics in Industry 4.0	Data analytics in IIoT optimizes industrial operations through real-time data analysis, fostering smart factories and seamless connectivity.
Network Flexibility Software-defined Networking (SDN)	SDN enhances network flexibility, enabling dynamic control and optimization of industrial networks.
Predictive Maintenance in Manufacturing	Predictive maintenance reduces downtime and enhances efficiency by anticipating equipment failures based on data analysis.
Supply Chain Optimization	Supply chain optimization through data analytics and machine learning improves productivity and responsiveness.
Artificial Intelligence Overview	AI and ML applications in industrial contexts enhance decision-making, automation, and efficiency.
Big Data	Big data analytics drives insights and informed decision-making, addressing challenges and leveraging future trends.

Topic	Keynotes
Introduction to R and Julia Programming Languages	R and Julia facilitate data analysis and modelling in IIoT projects, enhancing data-driven decision-making.
Overview of Hadoop in the Context of IIoT	Hadoop efficiently manages and processes large datasets from IIoT devices, supporting scalable data management and analytics.

Summary:

The chapter delves into the crucial role of big data analytics and software-defined networks (SDN) in driving Industry 4.0 advancements. It emphasizes the importance of data analytics in industrial applications, particularly within the Industrial Internet of Things (IIoT), for optimizing operations through real-time data analysis. SDN's contribution to network flexibility is explored, alongside predictive maintenance strategies, supply chain optimization techniques, and the integration of artificial intelligence (AI) and machine learning (ML) in industrial processes. Key topics covered include IIoT components, SDN fundamentals, predictive maintenance, supply chain optimization, AI's cross-sector applications, big data's role in business, associated challenges, and future trends. Additionally, introductory knowledge of programming languages like R and Julia for IIoT projects, and the significance of Hadoop in managing IIoT-generated data, are discussed.

Trivia Challenge: Industry 4.0

Welcome to the Industry 4.0 Trivia Challenge! Get ready to test your knowledge of the wacky world of Industry 4.0 with these hilarious multiple-choice questions. Each question is paired with some side-splitting facts and figures about Industry 4.0. Let us dive in and have a giggle!

1) How does Big Data Analytics and Software-Defined Networks (SDN) intersect in Industry 4.0?

    A) By throwing a party where data wears virtual networking hats.

    B) By exploring network flexibility through SDN and diving into data analytics.

    C) By creating a virtual reality where data and networks dance together.

    D) By transforming data into network superheroes with SDN capes.

2) **What role does data analytics play in Industry 4.0, according to the chapter?**

    A) Data becomes the DJ at the Industry 4.0 networking party.

    B) Data analytics helps optimize processes like predictive maintenance and supply chain.

    C) Data takes centre stage, performing stand-up comedy for networking hardware.

    D) Data becomes the ultimate networking guru, teaching SDN the art of flexibility.

3) **What does the chapter highlight about Software-Defined Networks (SDN)?**

    A) They're like traditional networks but with virtual party hats.

    B) They provide network flexibility and are crucial in Industry 4.0.

    C) They're the secret sauce that adds flavour to data analytics.

    D) They're the virtual playground where data analytics algorithms frolic.

4) **How does the chapter describe the importance of information pre-processing?**

    A) It's like serving appetizers before the big data feast.

    B) It's crucial for predictive maintenance and supply chain optimization applications.

    C) It's the warm-up routine before data analytics hits the networking dance floor.

    D) the secret ingredient makes data analytics algorithms perform magic tricks.

5) **What insights do learners gain regarding building models for IIoT applications?**

    A) They learn to teach IIoT devices how to salsa dance.

    B) They explore the world of artificial intelligence and big data technologies.

    C) They become IIoT architects, designing virtual skyscrapers of data.

    D) They train IIoT devices to tell jokes and entertain networking hardware.

6) **According to the chapter, which programming languages are learners introduced to in the context of IIoT?**

   A) Java and Python for teaching IIoT devices how to speak human.

   B) R and Julia for diving deeper into data analytics.

   C) C++ and Ruby for coding virtual networking parties.

   D) JavaScript and PHP for making IIoT devices sing karaoke.

7) **What does the chapter overview regard Hadoop in the context of IIoT?**

   A) How Hadoop throws virtual parties for IIoT devices.

   B) How Hadoop is used for big data processing and storage in IIoT applications.

   C) How Hadoop teaches IIoT devices to perform magic tricks.

   D) How Hadoop transforms IIoT data into virtual networking superheroes.

8) **What metaphor does the chapter describe the intersection of Big Data Analytics and SDN?**

   A) Like a dynamic duo, they team up to optimize Industry 4.0 processes.

   B) Like two peas in a pod, they share a virtual dance floor of data.

   C) Like a secret recipe, they blend to create the perfect networking dish.

   D) Like a virtual playground, they provide endless fun for IIoT devices.

9) **What adjective does the chapter use to describe the importance of SDN in Industry 4.0?**

   A) Crucial.

   B) Dynamic.

   C) Mysterious.

   D) Hilarious.

10) **What role does information pre-processing play in optimizing processes, according to the chapter?**

   A) It's like adding seasoning to a networking soup.

   B) It's crucial for predictive maintenance and supply chain optimization applications.

   C) It's the warm-up routine before data analytics hits the networking dance floor.

   D) It's the secret ingredient that makes IIoT devices sing karaoke.

Chapter 10

Integration of Different Technologies - Industry 4.0

Learning Objectives:

- Understand the advantages and importance of integrating transformative technologies such as AR/VR, AI, big data, and advanced analytics in the context of Industry 4.0.

- Explore how these technologies reshape industries and enhance business processes across various sectors.

- Learn about the interaction between AR/VR, AI, big data, and advanced analytics, and their collaborative applications in different domains.

- Recognize the challenges and opportunities associated with the integration of these technologies, including interoperability issues and complexity.

- Gain insights into the potential impacts of AR/VR, AI, big data, and advanced analytics on businesses and consumers.

- Understand the continuous technological progress in these domains and anticipate future developments and innovations.

10.1 Advantages of Integration

In the dynamic environment of the 21st century, transformative technologies have become powerful catalysts that reshape the way we interact with the world. Augmented and virtual reality (AR/VR), artificial intelligence (AI), big data and advanced analytics represent a leader in innovation that promises to transform business and human experience.

These technologies are more than new and are integrating into our daily lives, affecting the way we work, communicate and see reality. Augmented reality overlays digital data over the physical world and blurs the line between virtuality and reality.

Virtual reality offers an unprecedented experience by immersing people in a completely digital environment. Artificial intelligence mimics human intelligence, allowing machines to learn, reason and solve problems. Big data leverages large amounts of data to gain insights and make informed decisions. Advanced analytics, including data mining and business intelligence, help organisations derive intelligence from complex data.

10.1.1 The Importance of Reshaping Industries

The impact of technological change goes far beyond innovation, it plays an important role in the development of any business. From healthcare to manufacturing, entertainment to education, these technologies are transforming processes, increasing efficiency and driving innovation.

In healthcare, AI aids diagnoses, AR/VR transforms the patient experience, and big data supports personalised treatment plans. In production, smart factories use AR/VR for training and maintenance, while artificial intelligence improves production processes. Education uses AR/VR to realise optimal learning experiences, self-learning of skills, and data-driven understanding of big data.

In addition, these technologies are reshaping the human experience, increasing entertainment through competitive games and virtual entertainment, and bringing the same people together from long distances. Their importance lies not only in collaborations, but also in the integration of technologies by creating synergies that make them effective.

10.1.2 Integration as an important aspect of technology

The expression of technological change is integration, not isolation. Augmented reality seamlessly integrates content into the physical world to enhance the experience of the world. Artificial intelligence improves decision-making capabilities through machine learning and continuous adaptation to new information. The power of big data lies in the intersection of insights from multiple sources enabling better decisions. It weaves a narrative of understanding through advanced analysis, data exploration, and visualisation.

When reviewing the previous sections, it is important to know the interaction between these technologies. The lines between AR/VR, AI, big data and advanced analytics are blurring, creating new possibilities and challenges. This combination is becoming the backbone of the technology landscape, pushing us towards an era where innovation is not just a buzzword but a way of life.

In search of revolutionary technology, we are beginning to walk towards the future, where the lines between the real and the virtual are increasingly disappearing, where intelligence is man-made but has a far-reaching impact, and previous knowledge has shown that it sheds light on the challenges in today's world. These technologies are coming together to create a wave of innovation that promises to transform businesses, redefine experience, and elevate people to unprecedented heights.

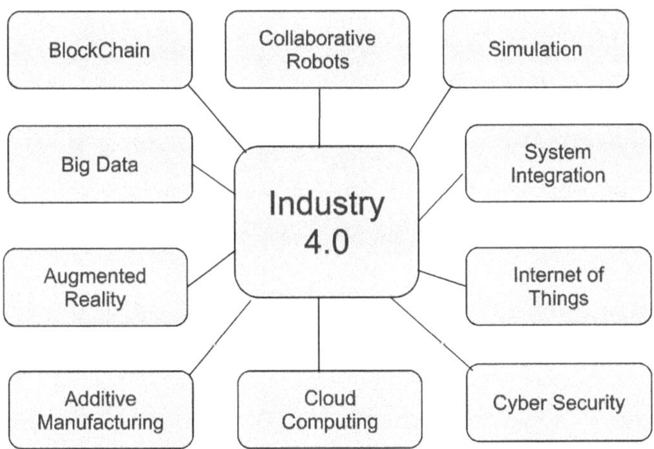

Exhibit 10.1 Industry 4.0 Elements

10.2 How do AR/VR, AI, big data, and advanced analytics interact?

The integration of augmented reality (AR), virtual reality (VR), artificial intelligence (AI), big data, and advanced analytics marks the era of Change. These technologies work together to create unique capabilities.

10.2.1 AR/VR and AI integration:

Improved user experience: AI-driven personalization features support the AR/VR interface, providing users with a better experience and adjusting accordingly.

Real-time adaptability: The artificial intelligence algorithm in the AR/VR system can adjust and respond in time according to the user's interaction, creating a good environment and operation.

10.2.2 Artificial Intelligence and Big Data:

Data-Driven Decision Making: Artificial Intelligence uses big data analytics to make data-driven predictions and decisions by obtaining valuable information from large data sets.

Improve intelligence model: Big data increases intelligence by providing diverse and detailed data to train machine learning models, increasing their accuracy and precision. yield.

10.2.3 Big Data and Advanced Analysis:

Innovative Data: Big data is the foundation of advanced analysis and provides the amount and type of data required for Check hard.

Meet predictive models: Advanced analytics techniques, such as predictive analytics, leverage rich data created and processed by big data.

10.2.4 Collaboration in business

Smart manufacturing:

AR/VR for training and supervision: AR/VR technology improves training plans and monitors the process in intelligent design, improving workers' skills and reducing downtime.

AI-driven predictive maintenance: AI analyses data from IoT sensors (generated from big data) to predict equipment failure, improve maintenance plans to improve and minimise production downtime.

Healthcare Diagnostics:

AR in Surgical Navigation: AR assists surgeons with precise navigation during surgery by combining real-time patient data (big data) with intelligence-based insights to improve outcomes.

AI-Assisted Diagnostic Tools: Artificial Intelligence algorithms analyse medical images (big data) to help faster and more accurate diagnosis, and AR/VR provides visual interaction for doctors.

10.3 Challenges and Opportunities in Collaboration

10.3.1 Interoperability:

Standardisation Efforts: Integration of these technologies faces collaboration challenges. Standardisation efforts are important to create seamless communication and compatibility.

Unified protocol: Create a unified data exchange and communication protocol to ensure AR/VR, AI, big data and analytics printing can work well together.

10.3.2 Using Combination Analyses:

Integration Complexity: Integrating multiple technologies requires addressing the complexity of different data and design processes.

Innovative Portfolio Insights: Overcoming integration challenges can unlock the potential of portfolio insights and lead to new applications and solutions across businesses.

As the industry increasingly integrates AR, VR, AI, big data and advanced analytics, integrated applications show potential for change. From smart manufacturing to health diagnostics, the combination of these technologies is more than the sum of its parts but it is the combination that drives business into new areas of efficiency, innovation and improved decision-making. Despite the challenges, the opportunities presented by the integration of these technologies promise to shape the future of business and the human experience.

10.3.3 How do these elements relate to each other?

AR, VR, AI, big data and advanced analytics are all interconnected. For example, AR and VR can visualise big data and provide insights that cannot be seen using traditional data visualisation tools. Artificial intelligence can be used to create new AR and VR applications. Big data and advanced analytics can be used to improve the performance of artificial intelligence.

10.3.4 Impacts on Businesses and Consumers

AR, VR, AI, big data and advanced analytics have many impacts on enterprise pressure and the customer.

Businesses can use this technology to improve efficiency, productivity and customer service. For example, companies can use AR and VR to train their employees, provide customer support, and market their products and services. Businesses can also use artificial intelligence and big data to improve decision-making and create new products and services.

Consumers can use this technology to improve their quality of life. For example, consumers can use AR and VR to play games, watch movies, and learn new things. Customers can also use artificial intelligence and big data to get personalised recommendations for products and services.

10.3.5 Challenges and Opportunities

AR, VR, artificial intelligence, big data and advanced analytics bring many challenges and opportunities.

Challenges include:

- The need to invest in new technology
- The requirement for employees to receive new skill training
- Cybersecurity threats
- Ethical competition

Opportunities include:

- Improved productivity

- Lower costs

- Advanced quality and efficiency innovation

- Greater agility and flexibility

AR, VR, AI, big data and advanced analytics are changing the way we live and work. They are powerful technologies. These technologies can transform many industries and improve consumers' quality of life. However, businesses and consumers need to be aware of the problems associated with this technology and take steps to alleviate them.

In conclusion, Augmented Reality (AR) and Virtual Reality (VR) are two new technologies that are rapidly changing the way we interact with the world around us.

AR is a technology that reflects digital information into the real world. It can be done through various devices such as smartphones, tablets and headphones. AR can be used for many applications such as gaming, navigation, and education.

VR is a technology that creates a fully simulated environment. This area can be covered with various devices such as headphones and gloves. VR can be used for many applications such as gaming, entertainment and education.

AR and VR can revolutionise many industries, including healthcare, manufacturing and education. For example, AR can provide doctors with real-time information during surgery. VR is used to train employees on new equipment or procedures. AR and VR can create a great learning experience for students.

10.3.6 Artificial Intelligence

Artificial Intelligence (AI) is a branch of computer science that aims to create intelligent agents or machines that can think, learn and act independently. Artificial intelligence; It can be used in many applications such as machine learning, natural language processing and computer vision.

Machine learning is a type of artificial intelligence that allows machines to learn from unstructured data. Natural language processing is artificial intelligence that enables machines to interpret and produce human language. Computer vision is a type of artificial intelligence that enables the machine to process and evaluate images and videos.

Numerous industries, including healthcare, banking, and transportation, have seen a transformation because of artificial intelligence. For example, artificial intelligence is used to develop new drugs and treatments for diseases. Artificial intelligence is also being used to create new products and financial services. Artificial intelligence is being used in the development of self-driving cars and trucks.

10.3.7 Big Data and Advanced Analytics

Big data refers to large and complex data that cannot be processed using the correct method. The process of providing relevant insights from large amounts of data is called advanced analytics.

Sales, manufacturing and healthcare are among the industries using big data and advanced analytics. For example, retailers are using big data and analytics to better understand customer behaviour and predict future demand. Manufacturers are using big data and analytics to improve production processes and reduce expenses. Doctors are using big data and clinical studies to improve patient care and develop new drugs.

10.3.8 Technological Progress Review

In our research on transformative technologies (augmented reality (AR), virtual reality (VR), artificial intelligence (AI), big data and advanced analytics), we have seen the evolution of industries into dynamic and interconnected ecosystems. From the experiences brought by AR/VR to the predictive power of AI and insights from massive data and advanced scanning, these technologies have revolutionised the way we work, communicate and innovate.

The integration of AR/VR and AI not only improves the user experience, but also opens up new possibilities for personal interaction and response. The combination of AI and big data unlocks the potential for success by allowing organisations to make informed decisions based on predictive analytics. Advanced analytics powered by big data has become the foundation of business, enabling actionable insights, informed decisions, and deep understanding of complex data.

Hope for Further Development

It is clear that the journey of technological development is new. The continued development of this technology is expected to further blur the lines between the physical and digital worlds. The integration of AR/VR, artificial intelligence, big data and advanced analytics will lead to unprecedented innovations and spur progress in many areas.

The way forward promises:

The way to drive better experiences: AI-powered AR/VR interfaces will continue to offer users a multitude of personalities and experiences tailored to their interests.

Data-driven insights: The collaboration between artificial intelligence and big data will deepen and lead to the emergence of more complex analytical models that can reveal complex patterns in data.

Industrial Innovation: Collaboration in smart manufacturing, diagnostics and other industries will flourish, creating new standards for business, safety and quality.

The effort to collaborate will prevail, technology will drive collaboration and enable the alignment of AR/VR, artificial intelligence, big data and advanced analytics skills.

Regarding Ethical Issues: With continuous efforts, technology ethics will remain at the forefront to address injustice, prevent privacy, and support responsible artificial intelligence and data practices.

The outcome of this research is not the end, but a stepping stone towards the future, where the integration of technological change will create a never-before-seen economy, pressure and human experience. The constant cycle of innovation requires us to be alert, flexible and ready to embrace opportunities in this technological field. As businesses and individuals embark on this path, the hope of innovation continues to guide the way and propel us into a future that continues to push boundaries.

Keynotes

Topic	Keynotes
Advantages of Integration	- Transformation of business processes - Reshaping industries - Enhancing human experience
Integration as an important aspect of technology	- AR/VR overlaying digital data - AI improving decision-making - Big data enabling insights
Interaction between AR/VR, AI, big data, and analytics	- Improved user experience in AR/VR - Data-driven decision making with AI and big data
Challenges and Opportunities in Collaboration	- Interoperability challenges - Complexity of integration - Potential for new applications

Summary:

This Chapter focuses on the integration of different technologies in Industry 4.0, particularly augmented reality (AR), virtual reality (VR), artificial intelligence (AI), big data, and advanced analytics. It highlights the advantages and importance of integrating these technologies, emphasizing their transformative impact on industries, business processes, and the human experience. The chapter explores how these technologies interact with each other and discusses their collaborative applications in various sectors such as healthcare, manufacturing, and education. Additionally, it addresses the challenges and opportunities associated with technology integration, including interoperability issues and ethical considerations. Overall, the chapter provides a comprehensive overview of the evolving technological landscape and its implications for businesses and consumers.

Trivia Challenge: Industry 4.0

Welcome to the Industry 4.0 Trivia Challenge! Get ready to test your knowledge of the wacky world of Industry 4.0 with these hilarious multiple-choice questions. Each question is paired with some side-splitting facts and figures about Industry 4.0. Let us dive in and have a giggle!

1) **What do you call it when AR, VR, AI, big data, and advanced analytics walk into a bar?**

   A) A tech-tastic party!

   B) The beginning of a revolutionary joke.

   C) Just another day in the world of Industry 4.0.

   D) The start of an integrated technological masterpiece!

2) **How do AR and VR technologies decide who gets the last data byte?**

   A) They engage in a virtual reality duel.

   B) They flip a digital coin.

   C) They consult with an AI mediator.

   D) They integrate their data-sharing protocols.

3) **Why did the big data analyst bring a ladder to work?**

   A) To reach the cloud computing servers.

   B) To climb the data hierarchy.

   C) To scale up the analytics.

   D) To get a better view of the integration landscape.

4) How do AI and big data unwind after a long day of collaboration?

> A) By running a relaxation algorithm.
>
> B) By binge-watching sci-fi movies.
>
> C) By analysing the patterns of their downtime.
>
> D) By integrating their leisure activities.

5) What do you get when you mix AR, VR, and AI in Industry 4.0?

> A) A virtual augmented intelligence extravaganza!
>
> B) The technological equivalent of a triple rainbow.
>
> C) A smarter, more immersive future.
>
> D) An integrated tech symphony.

6) Why did the AR headset break up with the VR goggles?

> A) They couldn't see eye-to-eye on integration strategies.
>
> B) They needed more space for their data.
>
> C) They wanted to explore other technological horizons.
>
> D) They realized they were better together.

7) How does big data handle stress in Industry 4.0?

> A) By compressing its worries into byte-sized chunks.
>
> B) By relying on AI for emotional support.
>
> C) By integrating mindfulness techniques into its algorithms.
>
> D) By finding patterns in chaos.

8) What's AR/VR's favourite genre of music?

> A) Techno-pop fusion.
>
> B) Virtual reality rock.
>
> C) Artificial intelligence hip-hop.
>
> D) Big data blues.

9) How do AI and advanced analytics settle disputes?

> A) By analysing the evidence.
>
> B) By integrating their decision-making processes.
>
> C) By flipping a digital coin.
>
> D) By running simulations of potential outcomes.

10) Why did the AI cross the data stream?

    A) To optimize its algorithms on the other side.

    B) To integrate with the information flow.

    C) To explore new patterns in uncharted territory.

    D) Because its programmer said so.

Chapter 11

Cyber Security - Industry 4.0

Learning Objectives:

- Understand the importance of cybersecurity in Industry 4.0.
- Identify the role of cybersecurity in protecting smart factories and connected systems.
- Recognize the various cybersecurity threats present in Industry 4.0 environments.
- Explore mitigation strategies and defence mechanisms against cyber threats.
- Understand the vulnerabilities inherent in Industry 4.0 systems.
- Recognize the significance of employee training in cybersecurity.
- Gain insight into the regulatory environment and compliance requirements in industrial cybersecurity.
- Explore emerging trends and future considerations in IIoT security.

11.0 Cyber Security

In the context of Industry 4.0, cyber security is gaining importance in protecting the digitalized ecosystems of networks and businesses. Industry 4.0 represents the fourth revolution, characterised by the always-there integration of smart technologies, automation and the Industrial Internet of Things (IIoT) into the production process. As business embraces digitalization and interconnection, its dependence on cyber-physical systems increases and requires strong cybersecurity measures.

11.0.1 Importance of Cyber Security

Protection of assets: Industry 4.0 involves the integration of physical and digital assets, making the protection of both physical and digital assets very important.

Continuous security: Cyber security is necessary to maintain uninterrupted operations and prevent disruptions caused by cyber threats or attacks.

Building trust: In the age of information exchange and collaboration, managing trust between stakeholders (such as customers and affiliates) relies on security and privacy assurances.

Reduce risk: As businesses become more interconnected, the scope of potential threats expands. Cyber security; It provides protection against risks such as data leakage, system control and identity theft.

Role in ensuring the security of smart factories and connected systems

Smart factories and connected systems are the backbone of today's business processes. Technologies such as the Internet of Things (IIoT), cloud computing and automation are used to increase efficiency and productivity in these areas. The role of cybersecurity in protecting this content is multifaceted.

Security Protection

Device Protection: Secure IoT devices, sensors, and connected systems through encryption and authentication protocols.

Data Integrity: Check and control the integrity of data created and processed in smart factories to prevent unauthorised access or modifications.

Ensure connected security:

End-to-end protection: Enable security measures across the entire value chain, from product to delivery, for end-user interactions.

Real-time threat detection: Continuously monitor and quickly detect network threats to the system to reduce response time.

Collaborative Security Culture: Promote a culture of cybersecurity awareness and collaboration among stakeholders, employees, and technology partners.

Incorporating cybersecurity into Industry 4.0 is critical to the sustainability and integrity of modern business. As smart factories and connected machines become part of the production process, taking strong cybersecurity measures is not only necessary but also crucial for sustainable operations and security in the era of Industry 4.0.

Type	Description
Network Security	Focuses on protecting the networking infrastructure, including devices, systems, and traffic within a network.
Application Security	Involves ensuring that all applications are securely developed, deployed, and maintained to prevent unauthorized access or breaches.

Type	Description
Cloud Security	Concerned with safeguarding data, applications, and infrastructure hosted in cloud environments from cyber threats and unauthorized access.
Internet Security	Specifically addresses threats that originate from the internet, such as malware, phishing, and other cyberattacks targeting web-based resources.
Endpoint Security	Involves securing individual devices like computers, smartphones, and tablets from various cyber threats, including malware, phishing, and unauthorized access.

Exhibit 11.1: Security Types

11.1 Introduction to the Cyber Security Threat Landscape in Industry 4.0

Definition Scope: The dynamic and interconnected nature of Industry 4.0 brings with it many cyber security threats.

Importance of Threats: Understanding threats is crucial to developing effective defence strategies.

Common cybersecurity threats:

Malware and ransomware: Examine the impact of malware and ransomware on business systems.

Phishing Attacks: Investigate techniques used to deceive employees and gain unauthorised access.

Insider Threat: Describes the risk posed by individuals within an organisation through violence or negligence.

Denial of Service (DoS) and Distributed Denial of Service (DDoS): Discuss the potential impact of excessive traffic causing a system to crash.

11.2 Objectives of Business Management

11.2.1 Understanding Industrial Control Systems (ICS)

ICS Overview: Examine the design and components of industrial control systems.

Role in Industry 4.0: Understand the critical role ICS plays in managing critical processes.

Targeted Attacks on ICS

Stuxnet and Other Attacks: Analyse the vulnerability of ICS infrastructure attacks, focusing on Stuxnet as a good example.

Zero-Day Vulnerabilities: Discover vulnerabilities that threaten exploits before developers fix them.

Supply Chain Attacks: Discuss the risks associated with supply chain interactions.

11.2.2 New threats in the age of digital transformation

Attack Report: Examining how the proliferation of technology expands the scope of cyber threats.

IT and OT Integration: Explore risks arising from the integration of information technology (IT) and operational technology (OT) environments.

Emerging Threats:

AI-Focused Threats: Learn how AI can be used to attack and defend.

IoT Exploits: Discuss the vulnerabilities created by the proliferation of Internet of Things (IoT) devices.

Cloud Security: Addressing problems and issues regarding the security of data stored and processed in the cloud environment.

11.2.3 Mitigation and Defence Strategies:

Effective Internal Defence:

Multi-Layered Approach: Emphasises the importance of security systems to create effective defences.

Risk Assessment and Management: Discuss appropriate procedures to identify and mitigate risks.

Collaborative Cybersecurity Culture:

Employee training: Emphasising the role of knowledgeable employees in reducing cybersecurity threats.

Discussion and Collaboration: Make a collaborative effort to share information among business partners.

Key Features Understand and mitigate cybersecurity threats in the Industry 4.0 environment. The chapter concludes by highlighting the need for greater attention, strategic reforms and collaboration to strengthen the resilience of business systems against cyber attacks.

11.3 Vulnerabilities in Industry 4.0 Systems

11.3.1 Vulnerabilities in Simple IoT Device

Lack of Security: Discuss the risks associated with vulnerability or lack of authentication in IoT devices.

Firmware and software vulnerabilities: Evaluate the challenges of managing and updating firmware and software across different IoT ecosystems.

Communication process insecurity: Solve the problem of IoT devices being easily and effectively compromised due to communication process insecurity.

11.3.2 Inadequate Cybersecurity

Cybersecurity in Industry 4.0

Characteristics of Interconnection: Explore the vast network that interconnects elements in Industry 4.0

The role of cybersecurity: Understand the role of critical and robust cybersecurity in protecting critical business processes.

11.3.3 Network security:

Inadequate access: Identify risks associated with inadequate encryption measures that could expose sensitive data to unauthorised access.

Vulnerability Management: Investigate the consequences of easy access that could allow unauthorised users to compromise network integrity.

Lack of Intrusion Detection Systems (IDS): Discuss the importance of actual threat detection and response processes.

11.3.4 Cloud-Based Vulnerabilities

Cloud Integration in Industry 4.0:

Leveraging Cloud Technologies: Understanding the benefits and challenges of integrating cloud services into Business 4.0.

Cloud-based data processing: Learn how cloud platforms process and process data.

Cloud-Based Security Vulnerabilities:

Privacy Information: Addresses risks associated with storing valuable information in the cloud.

Identity and Identity Access Management Challenges: Identify the challenges of managing user identity and access rights in cloud-based systems.

Risk Shared Services: Discuss the risks associated with sharing cloud resources in a multi-tenant environment.

11.3.5 Human Factors Contributing to Vulnerabilities

Human Factors in Industry 4.0 Security:

Employee Training and Awareness: Addresses the importance of cybersecurity training for those involved in Industry 4.0 gender roles.

Insider Threat: Discuss how to respond to intentional or unintentional breaches that could lead to security breaches.

Reducing human vulnerabilities:

Establishing a security protocol: Encouraging agreement on security and personnel procedures.

Continuing training plan: Provide regular training to raise employees' awareness of emerging cybersecurity risks.

Below shows the conflicts in industry 4.0 and highlights the need for a comprehensive and effective way to solve these problems. By addressing the vulnerabilities of IoT devices, strengthening network and cloud security, and identifying and mitigating human risks, businesses can leverage the impact of their systems to remediate cyber threats.

11.3.6 Cybersecurity Best Practices for Industry 4.0

11.3.6.1 Defence in Depth Strategies

Defining Defence in Depth: Consider the idea of using multiple strategies Protecting against different cyber threats layers of security to ensure.

Adaptation in Industry 4.0: Discuss dynamic defence strategies that evolve as the cybersecurity landscape changes.

Key Components of Defense in Depth:

Perimeter Security: Use strong walls and protection against outside access to your business network.

Endpoint security: Use anti-virus software, endpoint detection and response systems to protect the security of all devices.

Data security: To avoid unwanted access, encrypt critical information both in transit and at rest.

11.3.6.2 Encryption and Transport Security Protocols

Fundamentals of Transitioning to Industry 4.0:

Data Protection Integrity: Discussing how encryption ensures the confidentiality and integrity of business information.

Reducing the risk of eavesdropping: Exploring the role of encryption in preventing unauthorised disclosure of sensitive information.

11.3.6.3 Secure Communications:

TLS/SSL Protocols: Understand the importance of Transport Layer Security (TLS) and Secure Sockets Layer (SSL) in secure communications.

VPN Secure Connection: Discover how to use a virtual private network (VPN) to create a secure and private connection.

Real-time threat detection:

The need for continuous monitoring: Emphasising the importance of real-time monitoring to quickly identify and respond to potential threats.

Intrusion Detection Systems (IDS): Discussion of the role of IDS in actively monitoring networks for suspicious activity.

Effective incident response:

Incident Identification and Classification: Establishing protocols for recognizing and categorising cybersecurity incidents.

Rapid response mechanisms: Developing strategies for rapid response to incidents, minimising potential damage and downtime.

Exhibit 11.2 Cyber Security Verticals

11.4 The role of employee training in cybersecurity

11.4.1 Creating a cybersecurity culture:

Employee Education Programs: Outlining the importance of ongoing training to keep employees informed of evolving cyber threats.

Simulated Phishing Exercise: Conducting simulated phishing scenarios to educate employees on recognizing and avoiding phishing attempts.

11.4.2 Human-centred safety procedures:

Password Management: Educating employees on creating strong, unique passwords and the importance of regular updates.

Reporting protocols: Establish clear channels for reporting potential security incidents or concerns.

This chapter summarises the proactive cybersecurity measures essential to Industry 4.0 and highlights the need for a comprehensive approach. By implementing defence-in-depth strategies, using encryption and secure communications, ensuring continuous monitoring, and fostering a cybersecurity-aware culture through employee training, industries can increase their cybersecurity resilience in the dynamic Industry 4.0 environment.

11.5 Regulatory environment and compliance in industrial cyber security

11.5.1 Overview of Cyber Security Regulations in Industry 4.0

The evolving legal landscape: A discussion on how governments around the world are adapting to the challenges posed by Industry 4.0.

Sector-Specific Regulations: Examining Cyber Security Regulations Adapted to Sectors Strongly Impacted by Industry 4.0 Technologies.

11.5.2 Key Cyber Security Rules

GDPR in Europe: An analysis of the General Data Protection Regulation and its impact on personal data protection in Industry 4.0.

The NIST Framework in the United States: Understanding the National Institute of Standards and Technology Framework for Strengthening Critical Infrastructure Cybersecurity.

11.5.3 Compliance Challenges and Solutions

Cybersecurity Compliance Challenges:

Complexity of Industry 4.0 Systems: A discussion of the complex nature of interconnected cyber-physical systems and related compliance issues.

Rapid technological change: Addressing the difficulty of aligning compliance measures with the constant evolution of technology.

Compliance Solutions:

Continuous monitoring: Implementation of real-time monitoring systems to ensure ongoing compliance with regulatory standards.

Adaptive Compliance Strategies: Developing flexible compliance strategies capable of adapting to evolving regulatory requirements.

11.5.4 The role of international standards in enhancing cyber security

Standardisation in cyber security:

International Organization for Standardization: Emphasising the role of bodies such as ISO (International Organization for Standardization) in developing global cybersecurity standards.

Cross-Border Cooperation: A discussion on the importance of international cooperation in creating uniform cybersecurity standards.

The impact of standardisation on Industry 4.0:

Interoperability and compatibility: Exploring how standardised cybersecurity measures facilitate seamless integration and collaboration between different systems.

Risk Mitigation: A discussion on how adherence to international standards helps industries effectively mitigate cybersecurity risks.

11.5.5 Future trends in industrial cyber security

The role of artificial intelligence in cyber security:

Threat Detection with Artificial Intelligence: Exploring How Machine Learning and Artificial Intelligence Improve Cyber Threat Identification and Response.

Automated Incident Response: Discussing the potential of artificial intelligence in automating responses to cybersecurity incidents.

Blockchain and its potential applications:

Decentralised security: An analysis of how blockchain technology can contribute to the creation of decentralised and tamper-proof security systems.

Smart Contracts for Security: Exploring the Use of Blockchain-Based Smart Contracts in Ensuring Secure and Transparent Transactions.

Integrating cyber security into the design of Industry 4.0 systems:

Secure-by-Design principles: Promoting the incorporation of cyber security measures from the initial design phase of Industry 4.0 systems.

Cybersecurity as a Core Design Element: A discussion of the paradigm shift toward understanding cybersecurity as an integral part of the system architecture rather than an add-on feature.

Regulatory challenges, compliance strategies, the role of international standards and future trends in industrial cyber security – this chapter highlights the importance of a global and progressive approach. By complying with regulations, adopting standardised practices, and embracing new technologies such as AI and blockchain, industries can proactively address cybersecurity challenges in the dynamic Industry 4.0 environment.

11.5.6 The Dynamic Nature of Cyber Security Threats

Rapid Evolution of Threats: Exploring how cyber threats continue to evolve in complexity and sophistication.

Diversity of threats: Recognizing a diverse range of threats, from malware to social engineering, that target Industry 4.0 systems.

11.5.7 Evolving tactics of cyber adversaries:

Advanced Persistent Threats (APTs): Examining the persistent and targeted nature of cyber threats aimed at compromising industrial systems.

Social Engineering Techniques: Analysing Threat Actors' Manipulation of Human Psychology to Gain Unauthorised Access.

11.5.8 The Role of Threat Intelligence

Importance of threat intelligence:

Proactive Defence Strategies: A discussion of how timely and accurate threat intelligence enables proactive defence measures.

Collaborative Threat Intelligence Sharing: Emphasising the need for industry-wide collaboration in sharing threat intelligence.

Threat Intelligence Platforms:

Automating threat analysis: Exploring the use of automated platforms to analyse and disseminate threat intelligence.

Real-Time Threat Indicators: A discussion of how real-time indicators of compromise aid in immediate response to emerging threats.

11.6.8.1 Hygienic Cybersecurity Practices

11.5.8.1.1 Basic cybersecurity practices:

Regular software updates and patches: Emphasising the importance of keeping software and systems up-to-date to address known vulnerabilities.

Secure Configuration Management: A discussion of the importance of securely configuring systems from the outset.

Endpoint Security Precautions:

Endpoint Detection and Response (EDR): Exploring how EDR solutions improve endpoint detection and response capabilities.

User Privilege Management: Limiting user privileges to minimise the impact of potential breaches.

11.5.8.1.2 Building a Cyber Security Culture

Employee awareness and training:

Ongoing Training Programs: Advocating for ongoing cybersecurity training to keep employees informed of emerging threats.

Simulated Phishing Exercises: Running regular simulations to test and improve employees' ability to spot phishing attempts.

Cultivating a Security-Aware Workforce:

Encouraging whistleblowing: Creating a culture that encourages employees to immediately report any suspicious activity or security issues.

Recognizing and rewarding vigilance: Recognizing and rewarding employees for their proactive contributions to cybersecurity.

11.5.8.1.3 Adaptive Security Measures

Dynamic risk assessment:

Continuous Risk Assessment: Advocating for continuous cybersecurity risk assessment in the context of evolving threats and technologies.

Adaptive security policies: Developing flexible and adaptive security policies capable of responding to changing circumstances.

Technological innovations in adaptive security:

Artificial Intelligence in Threat Prediction: Exploring How AI-Driven Algorithms Can Anticipate and Adapt to Emerging Threats.

Behavioural Analytics: Using behavioural analytics to detect anomalies and potential security incidents.

This chapter emphasises the need for constant vigilance and adaptation and highlights the dynamic nature of cyber security threats in an Industry 4.0 environment. By fostering a cybersecurity-aware culture, leveraging threat intelligence, practising good cybersecurity hygiene, and adopting adaptive security measures, industries can strengthen their defences against evolving threats and ensure the ongoing resilience of their systems.

The Internet of Things (IIoT) is rapidly changing the business landscape, allowing businesses to increase efficiency, productivity and security.

But the Industrial Internet of Things also brings new security challenges. As more and more IIoT devices are deployed in the business environment, the parking space continues to expand and attackers are developing new methods and techniques to exploit these disadvantages.

11.5.8.1.4 Importance of Security in IIoT Application

As the industry embraces these benefits in the Internet of Things business, the importance of security is also important. IIoT applications involve interconnecting critical systems, systems, and sensitive data. Ensuring the security of these interconnected systems is critical to preventing unauthorised access, data loss, and potential business interruptions. The consequences of a security breach in IIoT can affect financial loss, the safety of people and the integrity of business processes.

Key factors that highlight the importance of security in IIoT applications are:

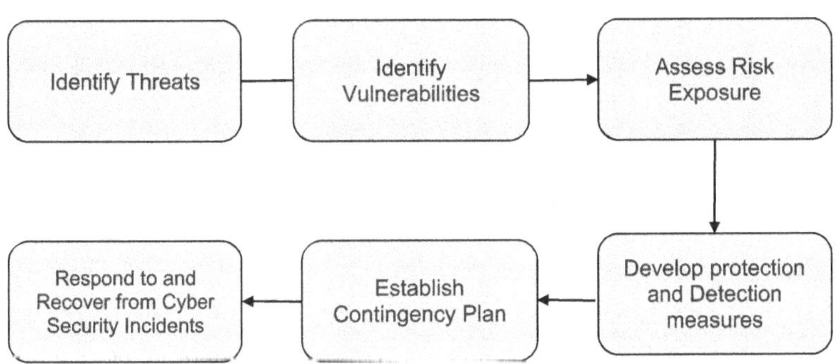

Exhibit 11.3 Threats Identification Areas

Data protection: Commercial IoT influences design and affects many information exchanges. Protecting this information against unauthorised access, interception or theft is crucial to maintaining confidentiality and business integrity.

Operations Continuity: Business processes are often critical and any disruption can have serious consequences. Security measures in IIoT are designed to reduce the risk of outages by ensuring that the communications network continues to operate efficiently.

Protection of Intellectual Property Rights: Many industries rely on systems and technologies. IIoT security protects intellectual property by preventing unauthorised access to data and trade secrets.

Safety Issues: In some industrial environments, IIoT applications have a direct impact on worker safety. Safety measures must be taken to prevent adverse effects that could harm people's health.

11.6 Changing threats in the business environment

Changing threats in the business environment require a strategy for security in IIoT applications.

These challenges include:

Cyber Security Threats: As business systems continue to evolve, the risk of cyber threats such as malware, ransomware, and phishing attacks increases. IIoT applications are a potential target for malicious actors trying to exploit vulnerabilities.

Interconnected System Vulnerabilities: Integration of legacy systems with IIoT technologies can create security vulnerabilities. Cyber attackers can exploit weaknesses in legacy systems to gain unauthorised access to the wider internet.

Supply Chain Risk: Industrial IoT often involves connected devices with multiple suppliers of products and services. Every point in the supply chain represents a potential entry point for cyber threats, making supply chain security a critical consideration.

Insider Threat: Malicious or unintentional actions of employees or contractors pose a serious threat. Insider threats range from negligence in underlying security procedures to deliberate actions that compromise the integrity of the IoT business.

As we move into the dynamic field of industrial IoT security, understanding these issues is crucial to implement effective measures to protect business and security. Comply with confidentiality, integrity and availability principles.

Current Challenges in IIoT Security:

The widespread use of the Industrial Internet of Things (IIoT) has introduced many connected devices and systems in technology and business. While these tools increase efficiency, they also bring significant security vulnerabilities.

Challenges: Many IIoT devices lack strong authentication mechanisms, making them vulnerable to unauthorised access. Weak or preset authentication combined with a lack of authentication mechanisms creates entry points for attackers.

Environment: Unauthorised access may lead to device settings being compromised, information leakage, or even public access to work.

Limited Security Updates and Patches:

Competition: Some IIoT devices may not receive security updates or patches. This is especially true for older products that were not designed with security in mind.

Impact: Unpatched vulnerabilities expose devices to known vulnerabilities, making them vulnerable to attacks that can disrupt the entire IIoT ecosystem.

11.7 Inadequate encryption techniques

Challenges: Inadequate use of encryption techniques makes IIoT data transmission vulnerable to interception and interception. Without strong encryption, sensitive data can be accessed by malicious actors.

Impact: Compromising data integrity compromises the confidentiality of critical information, including proprietary processes, trade secrets, and operational data.

11.7.1 No security boot mechanism:

Challenge: Most IIoT devices lack a security boot mechanism, making them vulnerable to firmware tampering. A malicious attacker could exploit this vulnerability to inject malicious code into the boot process.

Consequences: Compromising firmware integrity can lead to unauthorised control of the device, allowing an attacker to control functions or gain regular access to the IIoT network.

11.7.2 Interconnected Systems Vulnerabilities:

Challenges: IIoT environments often include new technologies along with existing legacy systems. These legacy systems may have outdated security systems, creating vulnerabilities in internet operations.

Environment: Cyber attackers can disrupt the entire IIoT ecosystem by exploiting weaknesses in legacy systems to gain access to critical systems.

Increasingly sophisticated cyber and Continuing advances in cyber threat sophistication characterise the evolving threat landscape in Industrial IoT security. The business environment faces increasing challenges as attackers become more adept at evading security measures.

11.7.2.1 Advanced Persistent Threats (APTs):

Challenge: APTs are targeted, persistent cyberattacks launched by well-funded organisations and advanced adversaries. These attacks often aim to access IIoT networks, extract sensitive data or disrupt business processes.

A side effect: APTs may go undetected for long periods of time and cause serious damage before being discovered. The persistence and success of these threats pose a major challenge to IIoT security.

11.7.2.2 Zero-Day Exploits:

Challenge: Zero-day attacks target an unknown vulnerability in software or hardware until developers release a patch. The rapid discovery and exploitation of these vulnerabilities poses an additional threat to IIoT security.

Impact: Zero-day vulnerabilities can be used by attackers to infiltrate IIoT networks, compromise devices, and perform malicious actions before security teams have a chance to respond.

Social Engineering and Phishing Attacks:

Challenges: Social engineering and phishing attacks often target human resources and use employees to gain unauthorised access, legitimise or delete sensitive information.

Environment: Successful social engineering can compromise credentials and allow attackers to use human resources as a vulnerable link to gain access to Industrial IoT networks and systems.

11.7.3 Ransomware and Extortion:

Challenge: The rise of ransomware assaults in mechanical situations includes scrambling basic information or frameworks, with assailants requesting instalment for unscrambling keys. Blackmail strategies are progressively utilised to disturb mechanical operations.

Impact: Ransomware assaults can end generation forms, causing budgetary misfortunes and operational downtime. The risk of blackmail includes a layer of complexity to IIoT security strategies.

11.7.4 Supply Chain Attacks:

Challenge: IIoT biological systems frequently depend on a complex supply chain with different sellers. Cyber aggressors target the supply chain to compromise gadgets or infuse pernicious components into the IIoT infrastructure.

Impact: Supply chain assaults can present compromised components, driving to vulnerabilities in IIoT gadgets. This strategy empowers aggressors to pick up unauthorised get to and possibly control basic systems.

Addressing the expanding modernity of cyber dangers requires a comprehensive and versatile approach to IIoT security. As innovation advances, security measures must ceaselessly progress to neutralise developing dangers and defend mechanical forms.

11.8 Emerging Trends in IIoT Security

11.8.1 Blockchain Integration

Ensuring Secure and Straightforward Transactions

Overview: Blockchain innovation is progressively coordinated into IIoT security systems to improve the security and straightforwardness of exchanges inside mechanical settings.

Implementation: Utilising blockchain for exchange confirmation guarantees that information trades between IIoT gadgets are secure, traceable, and safe to alter. Each exchange is recorded in a decentralised and unchanging record, lessening the chance of unauthorised alterations.

Benefits: This integration cultivates belief among partners by giving a straightforward and unquestionable record of all exchanges, upgrading the in general judgement of the IIoT ecosystem.

Immutable Records for Upgraded Information Integrity:

Overview: Blockchain's permanent record highlight guarantees that once information is recorded, it cannot be changed retroactively. This characteristic is utilised to improve information judgement inside IIoT environments.

Implementation: IIoT gadgets can record basic information, such as sensor readings or operational parameters, in a blockchain record. This guarantees that the data's realness and astuteness are kept up, anticipating unauthorised modifications.

Benefits: The permanence of blockchain records includes an additional layer of security, significant in businesses where precise and unaltered information is imperative for decision-making and compliance.

11.9 Manufactured Insights and Machine Learning

Predictive Analytics for Danger Detection:

Overview: Manufactured Insights (AI) and Machine Learning (ML) are tackled for prescient analytics to recognize and relieve potential security dangers some time recently they materialise.

Implementation: AI calculations analyse authentic information, gadget behaviours, and arrange designs to foresee potential security vulnerabilities or irregular exercises. This proactive approach empowers the pre-emptive tending of threats.

Benefits: Prescient analytics improves IIoT security by permitting organisations to expect and moderate dangers, diminishing the hazard of information breaches, unauthorised get to, and operational disruptions.

Adaptive Security Measures Based on Advancing Patterns:

Overview: AI and ML calculations powerfully adjust security measures based on advancing designs and patterns inside IIoT environments.

Implementation: Ceaseless checking of organised behaviours and irregularities permits AI frameworks to memorise and adjust security measures in genuine time. This adaptability is especially profitable within the confrontation of quickly advancing cyber threats.

Benefits: The versatile nature of AI-driven security measures guarantees that IIoT frameworks stay strong against developing dangers, giving a proactive defence component that advances with the changing risk landscape.

11.10 Edge Computing and Fog Computing: - IIoT Security

Addressing Idleness Challenges in Real-Time Processing:

Overview: Edge computing includes processing data closer to the source (IIoT gadgets) instead of depending on a centralised cloud framework. This addresses inactivity challenges and improves real-time preparing capabilities.

Implementation: Security conventions and measures are executed at the edge, guaranteeing that basic information is handled and analysed locally. This minimises the time it takes to distinguish and react to security incidents.

Benefits: By decreasing inactivity in information preparation, edge computing upgrades the responsiveness of IIoT security measures, pivotal in scenarios where quick activity is required to address potential threats.

11.11 Distributing Security Measures Closer to the Information Source

Overview: Mist computing expands the standards of edge computing by dispersing security measures all through the IIoT environment, making a fog-like layer that guarantees security is connected closer to the information source.

Implementation: Security measures such as encryption, confirmation, and peculiarity discovery are dispersed over edge gadgets, doors, and nearby servers. This dispersed approach fortifies the generally security pose of IIoT systems.

Benefits: Haze Computing upgrades the flexibility of IIoT security by decentralising security measures, making it more challenging for assailants to compromise the whole system.

11.12 Zero Trust Security Model - IIoT Security

Verification of Every Gadget and Client within the Network:

The Zero Believe security show works beneath the suspicion that no gadget or client, indeed those inside the organisation, ought to be trusted by default. Confirmation is required for each gadget and client endeavouring to get to resources.

Implementation: Persistent confirmation through instruments like multi-factor confirmation, gadget wellbeing checks, and character administration guarantees that as it were verified and authorised substances can get to IIoT resources.

Zero Believe minimises the assault surface, decreasing the hazard of unauthorised get to and horizontal development inside IIoT networks.

Continuous Observing and Authentication:

Persistent observing and confirmation are principal fundamentals of the Zero Believe demonstrate, guaranteeing that substances getting to IIoT frameworks are ceaselessly confirmed all through their interactions.

Real-time observing of client and gadget behaviours, coupled with versatile verification instruments, guarantees that any deviations from typical designs trigger prompt reactions, such as get to disavowal or increased security measures.

Nonstop checking upgrades the by and large security pose by recognizing and reacting to inconsistencies in genuine time, moderating the hazard of unauthorised get to or compromised credentials.

11.13 Quantum-Safe Cryptography

Preparing for Future Progressions in Quantum Computing:

Quantum computing postures a potential danger to conventional cryptographic calculations. Quantum-safe cryptography centres on creating encryption strategies that can withstand assaults from quantum computers.

Organisations are investigating and receiving quantum-resistant cryptographic calculations to future-proof their IIoT security. These calculations are planned to stand up to assaults from quantum computers, which have the potential to break routine encryption.

Quantum-safe cryptography guarantees that IIoT frameworks stay secure indeed in a future where quantum computing capabilities may debilitate the adequacy of existing cryptographic methods.

11.13.1 Implementing Cryptographic Calculations Safe to Quantum Attacks:

The usage of cryptographic calculations safe to quantum assaults includes embracing encryption methods that can withstand the computational control of quantum computers.

Organisations are transitioning to quantum-resistant calculations, such as lattice-based cryptography or hash-based cryptography, to secure communications and information inside IIoT environments.

By proactively executing quantum-safe cryptographic measures, IIoT frameworks are way better prepared to stand up to potential dangers posed by quantum computing advancements.

The integration of these rising patterns in IIoT security reflects a proactive reaction to the advancing danger scene. By leveraging innovations such as blockchain, AI, edge computing, the Zero Trust demonstration, and quantum-safe cryptography, organisations can upgrade the versatility of their IIoT biological systems and protect basic mechanical forms against progressively modern cyber dangers.

11.14 How to progress IIoT security in industrial settings

There are a few things that businesses can do to improve IIoT security: -

Implement a zero-trust security demonstration: A zero-trust security demonstration can offer assistance to ensure IIoT frameworks from assault by confirming the character of all gadgets and clients some time recently allowing access to resources.

Converge IT and OT security: Businesses must create a comprehensive security procedure that meets IT and OT security. It'll offer assistance to ensure all perspectives of an industrial operation from attack.

Use AI and ML for security: AI and ML can make strides in IIoT security in a few ways, such as identifying irregularities in organised activity, recognizing potential security dangers, and computerising security responses.

Focus on chance administration: Businesses must create a system that distinguishes and surveys the dangers related with their IIoT frameworks and actualizes fitting controls to relieve them.

By taking after these tips, businesses can move forward IIoT security in industrial settings and decrease the chance of attacks.

IIoT security could be a basic concern for businesses in all businesses. By understanding the key patterns in IIoT security and taking steps to make strides in their security pose, companies can ensure their IIoT frameworks from assault and guarantee the progression of their operations.

11.15 Success Stories in Securing Factory Environments

11.15.1 Integration of Bequest Frameworks at a Petrochemical Plant:

A petrochemical plant confronted challenges in securing its IIoT environment due to the integration of bequest control frameworks with present day IoT devices.

Segmentation and Separation: The plant executed division and confinement to secure basic bequest frameworks from potential cyber dangers beginning from IoT devices.

Gradual Modernization: Instead of a total upgrade, the plant received a slow modernization approach, overhauling bequest frameworks over time to adjust with present day security standards.

11.15.2 Automotive Fabrication Facility's Supply Chain Security:

A car fabricating office experienced security challenges in its IIoT supply chain, where different merchants were given components.

Vendor Security Appraisal: The office actualized a thorough seller security appraisal handle, guaranteeing that providers followed to indicated security standards.

Secure Communication Conventions: Actualizing secure communication conventions between the office and its providers made a difference to ensure touchy data amid information exchanges.

Continuous Checking: The office built up persistent observing homes to distinguish and react to any security issues inside the supply chain in genuine time.

11.15.3 Pharmaceutical Companies IIoT Gadget Security:

A pharmaceutical company confronted security challenges related to the multiplication of IIoT gadgets inside its fabrication processes.

Device Confirmation: Executing vigorous gadget confirmation conventions made a difference guarantee that as it were authorised IIoT gadgets might interface to the network.

Regular Security Reviews: Conducting customary security reviews and helplessness evaluations empowered the company to recognize and address potential shortcomings in its IIoT gadget ecosystem.

Employee Preparing: Comprehensive preparing programs were actualized to teach representatives on recognizing and reacting to potential security dangers, cultivating a culture of cybersecurity awareness.

These case considerations emphasise the significance of an all-encompassing and versatile approach to IIoT security. Victory stories highlight the adequacy of coordinated security arrangements, whereas challenges confronted give profitable experiences into the complexities of securing different mechanical situations. The lessons learned emphasise the importance of persistent checking, collaboration with industry accomplices, and proactive measures to remain ahead of advancing cyber dangers within the energetic scene of IIoT security.

11.15.4 Future Contemplations and Expected Improvements in IIoT Security

Advancing Danger Scene and Versatile Security Measures:

Dynamic Risk Landscape:

Anticipation: Long run of IIoT security will be formed by an ever-evolving risk scene. Rising cyber dangers, counting advanced malware, ransomware, and progressing tireless dangers, will proceed to target mechanical environments.

Adaptive Security Measures: To address these challenges, versatile security measures will be basic. The capacity to powerfully alter security conventions, risk discovery components, and occurrence reaction methodologies will be significant in relieving the effect of advancing threats.

Zero-Day Abuses and Risk Intelligence:

Zero-Day Abuses: Expecting the rise of zero-day abuses, organisations will contribute in progressing danger insights capabilities. Proactive checking and investigation of rising dangers will be basic to recognize and fix vulnerabilities some time recently they can be exploited.

Collaborative Danger Insights: Industry collaboration will play an essential part as organisations share threat insights to form a collective defence against novel and quickly advancing threats.

Human Component and Insider Threats:

Focus on Human-Centric Security: Recognizing the human component as a potential defencelessness, future IIoT security techniques will centre on human-centric security measures. Persistent worker preparing, mindfulness programs, and behavioural analytics will be utilised to relieve the chance of insider threats.

User Behaviour Analytics: Progressed client behaviour analytics will be coordinated to recognize atypical designs, empowering organisations to identify and react to potential insider dangers in genuine time.

Supply Chain Security:

End-to-End Supply Chain Security: With the expanding complexity of supply chains in IIoT environments, organisations will prioritise end-to-end supply chain security. This includes vetting and securing each point within the supply chain to anticipate compromise through the presentation of pernicious components or compromised devices.

Blockchain for Supply Chain Straightforwardness: The integration of blockchain innovation for upgraded straightforwardness and traceability in supply chains will end up broader, guaranteeing the astuteness of the complete IIoT supply chain

11.15.5 Proceeded Integration of Developing Advances in IIoT Security:

AI and Machine Learning Advancements:

Enhanced Prescient Analytics: AI and machine learning will see progressions in prescient analytics for indeed more exact risk discovery. These advances will progressively anticipate and react to security occurrences based on verifiable information, behavioural examination, and developing risk patterns.

Autonomous Security Operations: AI-driven security operations will become more independent, competent of making real-time choices and reactions without human mediation, moving forward the effectiveness of IIoT security measures.

Quantum-Safe Cryptography Standardization:

Widespread Selection: As quantum computing innovation propels, the appropriation of quantum-safe cryptography will be standardised over IIoT situations. Organisations will move to cryptographic calculations that can withstand quantum assaults, guaranteeing the long-term security of touchy data.

Integration with Existing Frameworks: Quantum-safe cryptography will be consistently coordinated into existing IIoT security systems, with a focus on keeping up in reverse compatibility and minimising disturbances to continuous operations.

Edge and Mist Computing Evolution:

Edge Security Development: Edge computing will advance with an increased centre on edge security development. Security measures, counting encryption, get to controls, and risk discovery, will be assisted fortified at the edge to ensure basic information handling points.

Decentralised Security Measures: Haze computing will witness the decentralisation of security measures, guaranteeing that security conventions are conveyed over the complete IIoT environment. This approach will improve the resilience of IIoT security by decreasing the effect of a single point of failure.

Blockchain for Improved Security:

Broader Application: Blockchain innovation will discover broader application past value-based security, expanding to gadget verification, secure firmware upgrades, and personality administration inside IIoT situations.

11.15.6 The role of cyber security in the Industry 4.0 environment

As the industry moves into the Industry 4.0 era, the role of cyber security has become more important than ever. Increasing connectivity and digitalization of business processes have created many cybersecurity challenges that must be addressed to ensure the integrity, confidentiality, and availability of important processes and information.

In Industry 4.0, cyber-physical systems are interconnected and the boundaries between physical and digital space are blurred. While this connectivity leads to unprecedented efficiency, it also exposes the business environment to cybersecurity threats. Perhaps the consequences of a cybersecurity breach in Industry 4.0 are far-reaching, including operational and financial losses, leakage of sensitive information, and even personal risk.

Cybersecurity is more than protection; It is an important component for safe and reliable business 4.0 technologies. By implementing cybersecurity measures, businesses can benefit from the benefits of digitalization without compromising the responsiveness and security of their operations.

In the next section, we will examine the challenges and solutions to cybersecurity in industry 4.0. By understanding the intersection of technology and business and the critical role cybersecurity plays, we can meet the challenges of this new business era with confidence and work.

11.15.7 The importance of cybersecurity in Industry 4.0

The emergence of Industry 4.0 Industry 4.0 has created a new era in the digitalization of connectivity and process work. As cyber-physical systems become an integral part of production and production, the importance of cybersecurity cannot be ignored. This chapter examines how cybersecurity has become an important part of Industry 4.0, the risks and challenges that arise in connected ecosystems, and the serious consequences of cybersecurity breaches in this time of change.

11.15.8 Cybersecurity Industry 4.0 as a Key Enabler of Industry 4.0

Cybersecurity is a key enabler in the context of Industry 4.0, where the integration of digital technology and industrial processes is ubiquitous. It promotes the integration of smart technologies and ensures the reliability and security of connected systems. Cybersecurity is more than a set of defences; It is an important component that enables businesses to gain self-confidence in benefiting from Industry 4.0 technologies.

In an environment where data is the lifeblood of operations, cybersecurity can protect sensitive data, assets and operational data. It promotes secure communication between devices, prevents unauthorised access, and ensures the confidentiality and integrity of critical processes. As businesses embrace automation, IoT devices, and real-time data analytics, the role of cybersecurity in ensuring the safe and efficient operation of these technologies becomes increasingly important.

11.15.9 Risks and challenges in the Connected Business Ecosystem

The connectivity of Industry 4.0 brings with it many risks and challenges that need to be addressed. Be careful. The features that made Industry 4.0 revolutionary—connected devices, instant data sharing, and autonomous decision-making—still create security vulnerabilities.

Key challenges include:

Multiple endpoints: As IoT devices and systems continue to connect, endpoints expand, providing more access points for cyber threats.

System Complexity: Cyber-Physical Complexity Systems in Industry 4.0 make it difficult to understand, monitor and protect the entire ecosystem.

Human factors: Human factors, including employees and contractors, represent the potential for security breaches through social engineering, insider threats, and lack of knowledge about cybersecurity.

Supply chain risk: As businesses become increasingly connected globally, the supply chain has become the target of cyber-attacks, posing risks to the integrity and availability of products and services.

Impact of cybersecurity breaches in Industry 4.0

The impact of cybersecurity breaches in Industry 4.0 goes beyond data breaches. They may include:

Operational disruptions: Cybersecurity issues can disrupt business processes, causing disruptions, delays, and financial losses.

Loss of Intellectual Property: Theft of property and personal information can harm a company's competitive advantage and innovation. surrounding environment.

Reputation Damage: Public trust is a valuable commodity. Cybersecurity breaches can cause reputational damage and undermine customer and stakeholder trust.

Legal and Regulatory: Failure to comply with cybersecurity regulations may result in legal action, fines, and audits.

Understanding the importance of cybersecurity In Industry 4.0, cybersecurity needs to recognize its role not only in protecting against cyber threats, but also in the safety and security of changing technologies. In this section, we will examine best practices, new technologies and strategies for the changing cybersecurity environment in Industry 4.0.

11.15.10 Cyber Security Threats in Industry 4.0

In the dynamic environment of Industry 4.0, the integration of digital technology into industrial processes will bring about a series of cyber security threats. This chapter provides an overview of cybersecurity threats in Industry 4.0, highlights the risks of attacks on business management, and explores the emerging threats in the age of digital transformation.

Exploring Cybersecurity Threats

As businesses engage in a combination of physical and digital operations, they are vulnerable to a variety of security threats, including

Malware: Poorly designed software In Industry 4.0, malware can target connected devices and thus affect the integrity and availability of critical systems.

Phishing attack: Fraud intended to obtain sensitive information by impersonating a trusted person. Phishing attacks can exploit human resources in business operations, resulting in unauthorised access to or deletion of information.

Denial of Service (DoS) and Distributed Denial of Service (DDoS) attacks: attempts to make a machine or network resource unusable by excessive users with high traffic. In Industry 4.0, DoS attacks can disrupt business processes and disrupt availability.

Insider Threat: The risk posed by people who have access to critical systems and information within an organisation. Insiders may intentionally or unintentionally compromise network security, causing information leaks or disruption of operations.

Targeted Attacks on Industrial Control Systems

One of the unique cybersecurity threats in Industry 4.0 is targeted attacks on Industrial Control Systems (ICS). Business management is an important part of important processes, production and business automation. Threats to ICS include:

Stuxnet attack class: Malware specifically designed to target and control programmable logic controllers (PLCs) and other components of ICS. One notorious example is Stuxnet, which was designed to disrupt Iran's nuclear program by targeting centrifuge control systems.

Zero-day vulnerability: An attack that does not involve an unexpected flaw in software or hardware. In the context of Industry 4.0, zero-day vulnerabilities are particularly important because they are not known to be protected.

Man, in the Middle (MitM) Attack: intercepts and controls communication between devices, which can lead to unauthorised or interrupted data flow in the control system.

New Threats in the Age of Digital Transformation

The digital transformation era of Industry 4.0 has introduced new and evolving cybersecurity threats, including:

Artificial intelligence-induced threats: artificial intelligence for execution and development purposes to increase the cyber-attacks. AI-driven threats can adapt and evolve on the fly, making detection and mitigation more difficult.

Supply chain attacks: Exploiting vulnerabilities in the supply chain for equipment, software or services used in Industry 4.0. A disruptive product can create hidden risks and vulnerabilities.

Ransomware attacks on business systems: Use ransomware to encrypt or block access to critical systems and demand payment for its release. In Industry 4.0, ransomware attacks can cause disruptions and financial loss.

Understanding these cybersecurity threats is crucial to building a strong defence in Industry 4.0. The following sections will highlight cybersecurity best practices, risk mitigation strategies, and the role of new technologies in protecting business processes in the digital age.

Vulnerabilities in IoT Devices

The growth of Internet of Things (IoT) devices is a sign of Industry 4.0, which brings more extremes to the business ecosystem. From sensors and actuators to smart machines, these devices all have certain vulnerabilities:

Insecure firmware: The firmware of many IoT devices may lack security protection. Insufficient firmware security may be used, resulting in unauthorised or unauthorised use of the device.

Weak authentication: IoT devices may have weak authentication mechanisms, making them vulnerable to unauthorised access. Barriers to accessing IoT devices could be a gateway to further attacks on the broader economy.

Lack of Access: Insufficient or no data encryption on Internet of Things devices can lead to sensitive data being compromised and subjected to unauthorised access. This vulnerability becomes serious when IoT devices process sensitive data in business environments.

11.16 Inadequate Network Security

The interaction of Industry 4.0 relies on networks that facilitate seamless communication between devices and systems. Inadequate network security leads to vulnerabilities that criminals can exploit: unauthorised access to critical systems. Segmentation is important to contain and isolate potential threats and prevent them from moving within the network.

Unencrypted communication: Unsecured communication in business networks allows information to be intercepted and compromised. The lack of encryption makes it easy for attackers to eavesdrop on communications between devices.

Weak controls: Weak controls can allow unauthorised users to access critical network components. Improving access control through strong authentication and authorization mechanisms is critical to network security.

Different level of Vulnerabilities

Cloud computing plays an important role in Industry 4.0 by providing scalability, flexibility and accessibility. But relying on cloud services has some drawbacks:

Insecure APIs: Application programming interfaces (APIs) that connect business systems to cloud services can be exploited without good protection. Unsecured APIs can reveal sensitive information or allow unauthorised access to cloud resources.

Data Privacy Issues: Storing valuable data in the cloud causes privacy issues. Inadequate data encryption and protection measures may expose sensitive data to unauthorised access or data leakage.

Trust in service providers: Businesses that rely heavily on cloud services may face security risks on their servers. Conflicts in cloud infrastructure can affect the security of business connections.

Human Factors Leading to Weaknesses

Despite advances in technology, human factors remain vulnerable to vulnerabilities in Industry 4.0:

Insufficient training: Inadequate cybersecurity training for employees increases the potential consequences of human error security events. Educating employees on cybersecurity best practices is critical to minimising security vulnerabilities.

Phishing and Social Engineering: People are susceptible to phishing attacks and social engineering tactics can lead to negative reactions. The knowledge base is essential to mitigate risks associated with human-centred vulnerabilities.

Device configuration errors: Security vulnerabilities can arise from user configuration errors during installation or maintenance. Proper training and documentation is essential to reduce the possibility of misconfiguration in an Industry 4.0 environment.

Solving this vulnerability requires an integrated approach that combines solutions, strong policies and regular training. In the next section, we will explain best practices in cybersecurity, risk mitigation strategies, and the role of new technologies in protecting Industry 4.0.

Cyber Security Best Practices for Industry 4.0

Security Environment Dynamic and Connected Environment The functioning of Industry 4.0 requires comprehensive training that includes assessment measures, good strategies and well-informed personnel. This chapter examines key cybersecurity best practices for Industry 4.0, including the critical role of defence in depth, encryption and secure communication protocols, continuous monitoring, incident response, and personnel training.

Defence in Depth-Strategy

According to changing cyber threats, defence in depth forms the basis of cyber security in industry 4.0. This approach involves implementing various security measures to create a protective barrier. Key components include:

Network Segmentation: Segment the business network to contain and isolate threats. Implementing network integration can limit attackers' movements and protect critical infrastructure.

Firewalls and Intrusion Detection/Prevention Systems (IDS/IPS): Use firewalls to monitor and control network access. Intrusion detection and prevention systems add an extra layer by instantly identifying and blocking malicious activity.

Access Control: Implement authentication and authorization mechanisms to restrict access to critical systems. Role-based management ensures that users have minimum permissions for their roles.

Endpoint Protection: Benefit from advanced endpoint protection that goes beyond traditional antivirus software. End-to-end protection should include capabilities such as behavioural analysis and intelligence response.

Access and communication security protocol

The use of encryption and connection security protocol is important to protect data integrity and confidentiality in Industry 4.0:

Data encryption: end-to-end for data in use, in transit and at rest -end-to-end encryption. This ensures that even if compromised, important data remains unread without the correct decryption key.

Communication security: Use industry-standard secure communications protocols such as HTTPS for networking and secure VPN for remote access. Use of security measures against unauthorised access and control of data.

Device Authentication: Use a strong device authentication mechanism to ensure only authorised devices can communicate. This prevents unauthorised devices from accessing critical systems.

Continuous monitoring and incident response

In the dynamic environment of Industry 4.0, continuous monitoring and incident response is crucial for threat detection and mitigation:

Real-time monitoring: Solutions with continuous monitoring Solutions that enable network activity, real-time visibility into vulnerabilities and security conditions. Automated monitoring tools enable timely analysis and response.

Incident Response Plan: Develop and regularly test an incident response plan to ensure a coordinated, rapid response to a safe incident. The plan should include steps for identification, retention, elimination, recovery, and lessons learned.

Threat Intelligence Integration: Accelerate the development of new network security threats by integrating intelligence sources. This enables protection measures based on existing threat areas and evolving vectors

The Significance of Employee Education in Cybersecurity

The human factor is still important in the cyber security equation. Employees are trained to be safe for the work 4.0 environment:

Cyber The importance of cybersecurity, best practices, and security compliance. Awareness can help reduce the risk of falling victim to social engineering attacks.

Phishing Simulation: Conduct a phishing simulation to test and improve employees' ability to identify and avoid phishing attempts. Simulation can reduce the risk of successful social engineering by supporting active and alert employees.

Device Security: Train employees in the use of security devices, including proper installation of devices, use of safe environment removal, and the importance of reporting security events. Check for suspicious activity immediately.

Organisations can strengthen their defences against threats in Industry 4.0 by following best practices in cybersecurity. The next section examines the regulatory environment, compliance challenges and new technologies that play an important role in improving the cybersecurity posture of Industry 4.0 systems.

Environmental Policy and Compliance Industry 4.0

With the digital transformation of Industry 4.0, many businesses are increasing their growth in the Industry Print 4.0 era and the management of the cyber security environment is increasing. This section examines the content of cybersecurity policies in Industry 4.0, the challenges organisations face in achieving this, and the role of international standards in improving cybersecurity.

Introducing Cybersecurity Regulations in Industry 4.0

Governments and regulators around the world are aware of the importance of cybersecurity in Industry 4.0. Many regulations and standards are designed to ensure the security, integrity and confidentiality of digital operations. Key elements of the regulatory environment include:

General Data Protection Regulation (GDPR): Although GDPR deals primarily with data protection and privacy, it also affects the security of business information. Organisations must comply with GDPR requirements when processing personal data in the context of Industry 4.0.

National and regional regulations: Many countries and regions have introduced special cyber security regulations for the business environment. These regulations outline the protection of critical infrastructure, the protection of sensitive data, and the reporting of cybersecurity incidents.

Specific business models: Some industries, such as healthcare, energy and finance, have specific cybersecurity business models and policies. These standards often include guidelines for business management and protection of sensitive information.

Compliance Challenges and Solutions

Ensuring compliance with cybersecurity regulations in the context of Industry 4.0 presents many challenges for stakeholders Compliance:

Multiple regulatory environments: Organisations' cross-border operations must comply with various requirements. Cyber security regulations. Meeting different needs is especially difficult for different companies.

Rapid progress: The rapid development of technology in Industry 4.0 may outpace the development of management systems. Keeping up with emerging threats and compliance requires speed and flexibility.

Limited capabilities: Ensuring compliance often requires significant investment in cybersecurity resources, including technology, technology personnel, and training. Limitations may impact the organisation's ability to meet compliance standards.

Solutions to these challenges include:

Global Intellectual Property Rights: Organizations must invest in intellectual property management to comply with international security regulations. Monitoring and evaluation studies help organisations plan and prepare for change.

Integrated Cybersecurity Framework: Implementing an integrated cybersecurity framework aligned to industry standards can lead to compliance. This process provides a way to meet regulatory requirements and manage cybersecurity risks.

Collaboration and knowledge sharing: Collaboration between organisations, trade associations, and regulators helps share best practices and consensus. Collaboration can help create a more unified approach to cybersecurity compliance.

11.17 The role of international standards in improving cybersecurity

International standards play an important role in harmonising cybersecurity practices and provide a framework for the international work of organisations:

ISO/IEC 27001: ISO/IEC 27001 standard It provides a framework for establishing, implementing, maintaining and continuously improving information security management systems (ISMS). It is widely recognized and adopted to improve cybersecurity practices.

NIST Cybersecurity Framework: Developed by the National Institute of Standards and Technology (NIST), this framework provides a risk-based approach to improving cybersecurity. It provides guidelines and best practices applicable to a variety of industries. IEC 62443: This international standard specifically addresses the security of industrial automation and control systems

(IACS). It provides a comprehensive guide to the use of cybersecurity measures in a business environment.

Adherence to international standards not only helps meet regulatory requirements, but also supports a cybersecurity posture that goes beyond regional and sectoral specificities.

In the next section, we will examine the continuous evolution of cybersecurity technologies and examine the use of these technologies in defence industry 4.0.

Future Trends in Industrial Cyber Security

As Industry 4.0 continues to evolve, the strategies and technologies used to protect businesses must also evolve. Let's explore key futures in the cybersecurity industry, focusing on the role of artificial intelligence (AI), the potential use of blockchain, and the integration of cybersecurity into business design. Edition 4.0 systems.

Function of Artificial Intelligence in Cybersecurity

Artificial Intelligence (AI) is emerging as a game-changer in the subject of industrial cybersecurity, providing superior talents to discover, analyse, and respond to cyber threats. Key factors of AI's function include:

Threat Detection and Analytics: AI-pushed analytics can sift via big datasets to pick out styles indicative of cyber threats. devices gaining knowledge of algorithms continuously enhance their detection abilities based totally on evolving threat landscapes.

Behavioural Evaluation: AI allows the monitoring of ordinary behaviour within commercial networks and structures. Deviations from hooked up baselines cause indicators, taking into account the fast identification of anomalies and ability threats.

Automatic Incident reaction: AI-driven incident reaction structures can automate the containment and remediation of cybersecurity incidents. This reduces reaction times and minimises the effect of cyber assaults on business operations.

Predictive Cybersecurity: AI models can predict capability vulnerabilities and proactively deal with them earlier than they're exploited. This shift from reactive to proactive cybersecurity enhances the resilience of commercial structures.

Blockchain and Its ability applications

Blockchain era, recognized for its decentralised in nature, holds promise for addressing particular cybersecurity challenges in industry 4.0:

Securing delivery Chains: Blockchain can beautify the safety and transparency of supply chains by presenting an immutable ledger for tracking the provenance of components and ensuring the integrity of products at some stage in the delivery chain.

Smart Contracts: Blockchain permits the usage of smart contracts, self-executing contracts with the terms of the agreement immediately written into code. In industrial settings, smart contracts can automate and at ease transactions, decreasing the hazard of fraud.

Data Integrity: The decentralised and tamper-resistant nature of blockchain makes it an appropriate technology for ensuring the integrity of critical information. In industry 4.0, wherein statistics accuracy is paramount, blockchain can offer a considered layer for information trade.

Identification and get admission to control: Blockchain-based totally identification answers can decorate the security of get entry to controls by way of supplying a decentralised and verifiable approach of authentication. that is particularly relevant in environments with numerous interconnected gadgets.

Integration of Cybersecurity into the layout of enterprise four.0 structures

An ahead-searching method to industrial cybersecurity includes integrating security features into the design and architecture of industry 4.0 systems:

Security with the aid of design: Embedding cybersecurity issues into the improvement procedure of business structures guarantees that security features are necessary instead of retrofitted. This approach minimises vulnerabilities from the outset.

Collaborative ecosystem protection: Industry 4.0 systems regularly involve collaboration across various stakeholders, including suppliers and companions. Designing protection functions that account for the entire ecosystem ensures a cohesive and complete defence method.

Human-Centric Safety: Spotting the position of human factors in cybersecurity, the layout of industry 4.0 systems should prioritise person-pleasant security interfaces, ongoing education programs, and focus campaigns to empower the group of workers towards cyber threats.

Adaptive security features: Designing systems with adaptive security features permits for actual-time adjustments based totally on evolving danger landscapes. This dynamic method complements the resilience of enterprise four.0 systems inside the face of emerging cyber threats.

As Industry 4.0 progresses, those developments in industrial cybersecurity will play a pivotal position in fortifying the resilience and integrity of interconnected systems. Within the following chapters, we can delve into actual-world case studies of successful IIoT implementations, demanding situations faced, and the destiny trajectory of IIoT in shaping industries.

Within the problematic panorama of Industry 4.0, wherein the convergence of technology and enterprise propels us into a new generation of interconnected structures, the crucial function of cybersecurity can't be overstated. This concluding chapter displays the necessary significance of cybersecurity in industry 4.0, emphasising the need for ongoing vigilance, model, and collective efforts to at ease the destiny of interconnected industries.

Through this research, we show that cybersecurity is important for Industry 4.0. As digital technologies are integrated into business processes, cybersecurity is used not only for protection but also as an important enabler. It protects sensitive data, ensures the integrity of critical systems, and enables businesses to leverage the transformative potential of interconnected cyber-physical systems.

From defence-in-depth techniques and encryption protocols to artificial intelligence and the combination of artificial intelligence and blockchain, cybersecurity is becoming the foundation for increasing the strength, reliability and trustworthiness of the Industry 4.0 environment. It is an ever-evolving shield against cyber threats that can disrupt operations, compromise security, and destroy stakeholder trust.

Adapting to this constant change requires vigilance and constant change

As Industry 4.0 continues, the scope of cybersecurity threats also increases. The ever-changing landscape of cyber dangers necessitates ongoing awareness and flexibility. Actors are constantly criticised and innovating, constantly adapting their ideas to take advantage of negative consequences. That's why organisations need to remain agile, proactive and educated to stay ahead of the ever-evolving threat landscape.

The use of technologies such as continuous monitoring, threat integration, artificial intelligence and blockchain is not only recommended but also important in countering persistent cyber threats. Investing in network security is an investment in the future security and prosperity of the Internet business.

For the future of Internet business after Industry 4.0, security is the brand's purpose and responsibility. It involves collaboration between industry, government, regulators and cybersecurity experts. It requires adherence to best practices, compliance with regulations, and a desire to improve prospects and reduce emerging threats.

As we move towards this time of change, the convergence of technology and business opens the way for new opportunities and efficiencies. However, this integration still requires a common commitment to cybersecurity as a fundamental principle. By applying the principles outlined in this research—whether by implementing a defence-in-depth strategy, AI-driven threat detection, blockchain-enhanced security, or incorporating cybersecurity into the design—we lay the foundation for security and pave the way to success. future.

Security for the future of business networking is not only about crime prevention, but also about developing and adapting a cybersecurity culture. It's about creating a mind-set that sees cybersecurity not as a static problem, but as a powerful force moving the security industry forward.

As we complete our journey towards Industry 4.0 and cybersecurity, the statement is clear, the future is interconnected and the security of the future is within us. The decisions we make today, the technology we acquire and the measures we take are not only a new and useful idea, but more importantly, the secure future of jobs will shape the environment.

Topic	Keynotes
Importance of cybersecurity in Industry 4.0	The importance of cybersecurity in Industry 4.0
Role of cybersecurity in smart factories and connected systems	Understanding the role of cybersecurity in smart factories and connected systems
Cybersecurity threats in Industry 4.0	Identifying cybersecurity threats in Industry 4.0
Mitigation strategies and defence mechanisms	Exploring mitigation strategies and defence mechanisms against cyber threats
Vulnerabilities in Industry 4.0 systems	Recognizing vulnerabilities in Industry 4.0 systems
Employee training in cybersecurity	Understanding the importance of employee training in cybersecurity

Topic	Keynotes
Regulatory environment and compliance in industrial cybersecurity	Overview of the regulatory environment and compliance requirements in industrial cybersecurity
Emerging trends in IIoT security	Exploring emerging trends in IIoT security

Summary:

Cybersecurity plays a crucial role in Industry 4.0, where the integration of digital technology and industrial processes is widespread. It ensures the reliability and security of connected systems, protects sensitive data and assets, and promotes secure communication between devices. However, Industry 4.0 environments face various cybersecurity threats, including ransomware, supply chain attacks, and social engineering. To address these challenges, businesses must implement mitigation strategies, prioritize employee training, and adhere to regulatory standards. Looking ahead, emerging trends such as blockchain integration and quantum-safe cryptography will shape the future of IIoT security.

Trivia Challenge: Industry 4.0

Welcome to the Industry 4.0 Trivia Challenge! Get ready to test your knowledge of the wacky world of Industry 4.0 with these hilarious multiple-choice questions. Each question is paired with some side-splitting facts and figures about Industry 4.0. Let us dive in and have a giggle!

1) Why did the cybersecurity expert bring a flashlight to work?

    A) To shed light on potential vulnerabilities.

    B) To illuminate the dark web.

    C) To brighten up the firewall.

    D) To navigate through the fog of cyber threats.

2) How do smart factories ensure they're safe from cyber-attacks?

    A) By installing antivirus software on assembly lines.

    B) By encrypting every bite of data, they produce.

    C) By integrating cybersecurity measures into their operations.

    D) By hiring cyber watchdogs to patrol the premises.

3) What's a hacker's favorited type of cloud?

A) The one where they store their stolen data.

B) The dark cloud looming over unsecured systems.

C) The cloud of uncertainty surrounding cyber defences.

D) The one they access from their virtual lair.

4) How do cyber threats navigate the IIoT landscape?

A) By exploiting vulnerabilities in edge computing.

B) By bypassing fog computing protocols.

C) By phishing for data in smart factory networks.

D) By integrating malware into IoT devices.

5) Why did the firewall blush?

A) Because it detected a breach in its defences.

B) Because it received a compliment from the network administrator.

C) Because it was overheating from blocking cyber-attacks.

D) Because it realized it forgot to encrypt its login credentials.

6) How do cyber defences stay ahead of emerging threats?

A) By integrating AI into their security protocols.

B) By disconnecting from the internet entirely.

C) By outsourcing cybersecurity to the cloud.

D) By building higher firewalls around their networks.

7) What do cybersecurity experts do when they're not fighting cybercrime?

A) They attend hacker conventions undercover.

B) They analyse the latest cyber threat trends.

C) They practice their digital karate moves.

D) They integrate new security patches into their systems.

8) Why did the IoT device go to cybersecurity school?

A) To learn how to encrypt its data transmissions.

B) To defend itself against cyber-attacks.

C) To integrate itself into secure networks.

D) To avoid becoming a target for hackers.

9) What's a cybercriminal's favourite holiday?

A) April Fools' Day, when they unleash their pranks.

B) Halloween, when they wear their digital masks.

C) Cyber Monday, when they target online shoppers.

D) Valentine's Day, when they steal hearts and data.

10) How do cyber defenders keep their cool under pressure?

A) By practicing their cyber defence drills.

B) By integrating stress-relief techniques into their protocols.

C) By relying on AI to detect and neutralize threats.

D) By understanding that cybersecurity is no joke.

Abbreviations

Abbreviation	Terms
IC4	Importance of cybersecurity in Industry 4.0
RC-SFCS	Role of cybersecurity in smart factories and connected systems
CT-I4	Cybersecurity threats in Industry 4.0
MS-DM	Mitigation strategies and defence mechanisms
V-I4S	Vulnerabilities in Industry 4.0 systems
ET-CS	Employee training in cybersecurity
REC-ICS	Regulatory environment and compliance in industrial cybersecurity
ET-IIOTS	Emerging trends in IIoT security

Chapter 12

Digital Strategy Framework – Industry 4.0

Learning Objectives:

- Framework for developing a digital strategy in the context of Industry 4.0.

- Various components, such as meeting customer expectations, governing entities' roles, and leveraging third-party networks.

- Critical internal forces driving Industry 4.0 integration, levels of adoption, and the necessity of a digital strategy for companies.

- Production automation, process reconstruction, and digital platform development.

- Explore challenges in adopting Industry 4.0 technologies and the roles of leadership, investors, and boards in digital transformation.

- The emergence of new roles like Chief Digital Officer, profile structure, reporting, and talent acquisition strategies.

12.0 Introduction

Digital strategy development requires consideration of competitive dynamics. Rival firms' actions within an industry significantly influence strategic planning. Competitors and industry conditions are important factors shaping digital strategy. To remain successful, companies must develop digital strategies to either gain an advantage over competitors or, in many cases, simply maintain parity.

While competition specifically relates to how peer businesses behave, industry growth and change more broadly impact overall market conditions. In other words, a company may reactively pursue a digital strategy based on its sector's current state. During times of disruption, firms must act with agility and strategic flexibility to endure. In periods of industry expansion, all participants are profitable and acquiring customers. The need for differentiation lessens during growth, and companies will digitally transform only as much as peers within the industrial culture.

12.1 Meeting Customer Expectations

Customers who utilise digital technologies in their personal lives expect similar solutions and strategies when engaging in business transactions. Companies are implementing digital strategies to supply customers with the right products at the optimal time and location to satisfy these expectations. In business-to-business transactions, customers themselves are undergoing digital transformation and require their suppliers and partners to synchronise accordingly. Digital technologies enhance customer expectations of suppliers and merchants. This includes availability of products online and across multiple channels, integrated supply chains and delivery, order tracking mechanisms, multi-channel and real-time reporting, and customer support. The sole manner a company can achieve all of these is by adopting digital strategies.

12.2 The Role of Governing Entities

Governing entities like private equity, venture capital and financial institutions serving on company boards play a role in challenging management to identify new growth and transformation opportunities while also approving major funding decisions. These governing entities play a significant part throughout the digital transformation process for start-ups and at scale. They are involved in encouraging companies to embark on their digital journey, providing financial resources, monitoring impact, offering strategic guidance, helping build networks and implementing course corrections. Governing entities have the advantage of involvement in the digital transformations of multiple portfolio companies across industries. They furnish company leadership and management with strategic insights as well as best practices from these multiple industries.

12.3 Leveraging Third-Party Networks

Most companies have started leveraging external platforms for sales, marketing and procurement of inputs. E-commerce platforms like Amazon and supply chain platforms such as Ariba and Alibaba represent some examples.

The Role of Government Support

To boost national economic growth and employment, governments are encouraging companies to adopt digital strategies. Interventions by central governments, government agencies and industry bodies are critical for digital adoption, especially among smaller companies. These include policy support, provision of funding and subsidies as well as digital skills training. Governments themselves are adopting digital technologies, and many countries are witnessing co-evolution of digital capabilities within government agencies and industry.

12.4 Key Internal Forces Steering Industry 4.0 Integration

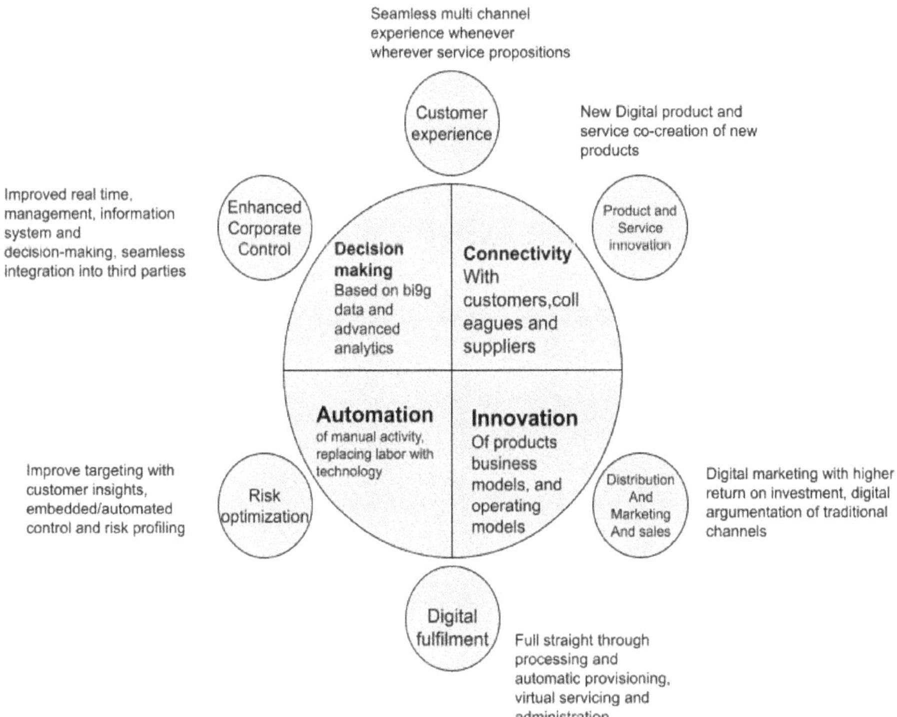

Exhibit 12.1 Digitisation workflow for Enterprises

12.4.1 Dynamic Leadership and Digital Strategy

Digital strategy is a high priority for senior leadership in many companies. Studies have shown that companies leading in digital transformation typically have visionary leadership focused on developing a strong digital strategy. As digital strategies involve multiple business functions, strong leadership is needed to coordinate efforts.

While the role of Chief Digital Officer continues to evolve in organisations, the CEO plays a critical role in initiating transformation. The CEO is responsible for setting overall direction, facilitating collaboration between IT and business teams, and governing the transformation process.

To set direction, CEOs must actively seek input. Traditionally, input was limited to a company's industry and related sectors. However, in today's dynamic environment, CEOs need an adaptive leadership approach that incorporates ongoing feedback from within the company to enable course corrections as needed. CEOs also need to cultivate a culture of open innovation where digital strategies are continuously evaluated.

Leadership age also influences digital innovation adoption. Companies led by Millennials tend to embrace new technologies more readily. Younger leaders incorporate digital concepts earlier due to personal experience with technology in business and personal life. In traditional family businesses, digital transformation often accelerates as Millennial generations take on larger roles. Companies founded by Millennials are born digital with digital strategies integrated from the start. Traditional companies may also transform digitally when they recognize competition from born-digital peers led by Millennials. The threat of falling behind drives transformation.

12.4.2 Growth and Customer Connectivity

A primary driver for digital strategies is increased customer connectivity and business growth. Digital enables connecting with customers through the right products, channels, and timing. For example, using data analytics and AI, a retailer can provide customised products to each customer at the optimal time. Omni channel strategies create seamless buying experiences. Digital also increases customer touchpoints, requiring sophisticated customer relationship management.

Digital transformation supports lean and flexible growth at unprecedented scales. Flexibility enables agile innovation. E-commerce platforms provide instant global market access. Digital technologies like IoT and RFID enable tracking physical goods movements, especially benefiting emerging markets with infrastructure gaps.

12.4.3 Turbocharge Process Efficiency

Most of the early-stage companies stated that supercharging efficiency and customer experience is the reason for their digital strategy. Digital technologies help in rocketing the efficiency of a company to new heights. They help in streamlining and integrating various parts of the value chain seamlessly for maximum efficiency.

Digital technologies also help in eliminating slowdowns in process workflow, automate manual interventions, provide awesome reporting and monitoring mechanisms, predict process defects by proactively monitoring and so much more. IoT, robotics and virtual reality (VR) are some important digital technologies that are implemented to powerfully enhance process efficiencies. The data that emerges from a digital implementation can be used to drive further process efficiencies to even greater levels.

Digital Strategies have helped in implementing technologies that integrate business processes across a company and its network partners like nothing else. As product life cycles shorten and R&D costs spiral out of control, concepts such as open innovation enabled by digital technologies are becoming more prevalent.

Industry 4.0 has been used to describe the digital transformation of production, and manufacturing environments where technologies such as CLOUD, BLOCKCHAIN, IoT, and Big Data are being integrated to build smart manufacturing like never before. Industry 4.0 brings value creation opportunities for all types of companies. For a Supply chain process, Industry 4.0, sometimes referred to as Procurement 4.0, provides unparalleled agility, dynamic co-operation, and ability to operate beyond organisational and national boundaries. Digital has allowed for companies to rapidly set up new supply chain networks and gain a first-mover advantage for new products unlike ever before. Digital transformation of supply chains has helped increase current efficiencies and enabled companies to build agile processes to develop, launch and market new products like never before.

At production level, increased productivity and flexibility allow for smaller batch sizes and automate production decisions, and big data and analytics combined with IoT help in proactively detecting errors and defects. This helps in increasing customer experience and customer satisfaction to new heights.

Implementing a standard digital strategy across an organisation, including ERP and supply chain management systems, can be complex and time-consuming. However, once implemented, it can lead to greater efficiency and agility unlike anything seen before.

12.4.4 Go Global and Explore

Going global and exploring new markets is one of the key triggers of a digital strategy. Digital enables instant cross-border trade unlike ever before. This trade encompasses a whole variety of transactions. E-Commerce platforms have enabled sourcing of materials from a global supplier base unlike ever before. On the other end, these platforms have allowed for companies to expand their markets outside of their national boundaries unlike ever before. In the pre-digital era, companies, especially small companies, did not have the resources to go beyond their regional markets. Digital has enabled them to access these markets with very minimal set-up costs and efforts and has reduced the time to do so unlike anything seen before.

Another aspect of digital is accessing global talent. While this has been primarily in the services sector, digital platforms have allowed companies to source talent globally unlike ever before.

12.4.5 Attracting Top Talent

Digital strategies have a significant impact on organisation' talent strategies. On one hand, digital platforms like artificial intelligence, chatbots and robotic process automation eliminate the need for manual work and large teams. As a result, companies will need lean workforces and procure many outsourced and contract employees. Digital platforms will be used to source personnel and teams as needed. On the other hand, learning functions must leverage digital tools to provide training on innovation, change management and agility in addition to skills for digital transformation. Specialised talent will be most sought after.

Networking and collaboration will be critical for fostering innovation and agility. Creating a digital workplace that incorporates technologies such as big data, cloud computing and AI can help enhance individual and organisational productivity.

While implementing a digital strategy requires certain skills, companies also use their digital maturity to attract top talent. Younger workers, especially millennials, want to feel they are part of the next big opportunity. Therefore, a company needs to project that it is a digitally mature organisation or on a path to digital transformation. The term 'digital' has also been added to job titles to attract the right talent.

12.5 Level of Adoption

Digital adoption exists on a continuum, and the line between digitization and digital transformation is fine. Most technology companies and start-ups are digital experts. On the other end of the spectrum, there are hardly any companies that are digitally unaware. The COVID-19 pandemic accelerated digital adoption. As a result, most companies are in the "digital aware," "starter" or "proficient" phases. The degree of digital adoption also varies by industry. Industries like financial services, healthcare and retail have seen greater adoption of digital technologies. Manufacturing, small and medium-sized enterprises, and agribusiness are in early stages of adoption.

Level	Definition
Unaware	Limited to no awareness of digital transformation or technology's impact on business operations.
Exploring	Developing basic awareness of digitization and how automation can affect processes. Potential opportunities and use cases are being explored.
Initial Implementation	At least one digital technology has been implemented, although digital strategies have yet to be fully integrated.
Engaged	A digital roadmap has been established. Multiple digital technologies are being deployed aligned with transforming the business model. Process optimization through technology is underway.
Digitally Native	The organisation operates with digital as the primary operating model. Technology is fully integrated across all functions to drive innovation, customer experience and business growth. Data and automation underpin all business operations.

Exhibit: 12.2 Digital Maturity Levels

12.6 Does Every Company Need a Digital Strategy?

Every CEO is so overwhelmed with digital info it's like they fell into an internet wormhole. They go to all these conferences and panels hoping to score the secret sauce to success, and all they hear is "disrupt or be disrupted!" Now while going digital has its perks, CEOs gotta be careful not to lose their marbles over it.

The basics of good biz strategy haven't changed, folks. You still need to know what your customers want, both now and in the future. Then you gotta see how your strengths match up against what's out there. A digital plan is just using tech tools to make your plan happen. Once you figure out your strategy, then you decide which techy toys will do the job.

Some options may involve going online, but others will be old school reliable. A few key questions to mull over: Does going digital fill a need customers have? Will it help you reach more people or market share? Will it help you stay ahead of the Joneses? Does it give you special new skills? And does digitising actually make financial sense?

The answers should show you how hard-core digital needs to be in your master plan. Just randomly adding gizmos won't get the job done - you got to use tech to advance your overall mission. Stay focused on the fundamentals, my dudes, and the pixels will follow!

12.7 Production Automation

Production automation is a type of process automation primarily used in manufacturing contexts. It is generally classified into three categories:

1. Fixed automation utilises dedicated equipment designed for specific purposes to handle high volumes and eliminate manual tasks. Conveyor belts connecting areas of a manufacturing plant exemplify fixed automation. With narrow application and low cost, fixed automation serves dedicated functions.

2. Programmable automation employs flexible equipment to control batch processes, like mixing chemicals according to a program in pharmaceutical production. Monitoring parameters including temperature and pressure, programmable automation automates lower-speed processes requiring intermittent adjustment at higher cost than fixed automation.

3. Flexible automation, a programmable automation variation, centralises control. Applicable to smaller batches and broader product ranges causing high downtime and waste with programmable automation, flexible automation utilises reconfigurable robots and workstations.

Digital technologies as IoT and robotics comprise key flexible automation elements. Tesla's Fremont, California plant demonstrates flexible automation. Tesla internally developed its Tesla Manufacturing Operating System allowing rapid reconfiguration of automation equipment, including robots, and faster improvements or new product development than traditional automakers.

Process	Description
Production Automation	Classification of production automation into fixed, programmable, and flexible categories - Examples and characteristics of each automation type - Integration of digital technologies in flexible automation
Process Reconstruction	Historical context of vertical integration in business - Reasons for the decline of vertical integration - Transition to outsourcing and offshoring strategies - Role of digital technologies in enabling outsourcing and offshoring - Unbundling of the value chain and its impact on business operations and innovation

Process	Description
Product Innovation	Types of innovation: incremental, radical, and disruptive - Examples and characteristics of each innovation type - The role of digital technologies in enabling product innovation - Necessity of New Product Development (NPD) process - Key steps and considerations in the NPD process - Market scanning, initial market analysis, and digital platform development as integral parts of product development process

Exhibit 12.3 Process & Product Innovation

Business process automation (BPA) similarly automates recurring business activities as in sales, supply chain, and finance/accounting. Examples include accounts payable and sales reporting automation. The next phase utilises robotics process automation (RPA) with digital tools like software robots and AI-based digital workers performing automated tasks by observing user actions. AI chatbots exemplify RPA's growing customer service use.

12.8 Process Reconstruction

Historically, companies were vertically integrated and owned all aspects of the value chain. Some reasons for vertical integration included: High transaction costs: Without modern technology and telecommunications, managing multiple suppliers across geographies was very expensive. Supply uncertainty: Integration eliminated dependencies on suppliers. Companies had full control over their cost structure and less risk of disruption in raw material supply. Raising barriers to entry: By controlling all critical elements in the supply chain, companies ensure no other competitor could enter the market. However, vertical integration is no longer a preferred strategy. Several factors have driven this shift. Vertically integrated companies became large and unwieldy. They lost their edge in innovation as costs grew due to management overhead required to run a complex conglomerate. Companies now use outsourcing, offshoring, and unbundling of value chains to reduce complexity and be more adaptable to meet customer needs.

12.8.1 Outsourcing and Offshoring Strategies

Outsourcing refers to the practice of contracting out part of a company's production or business processes to external specialist suppliers. As a strategy, outsourcing has been utilised for over 100 years. Ford Motor Company, an early pioneer in automobile manufacturing that led the second industrial revolution, originally maintained vertical integration across its entire value chain. In the early 1920s, Ford controlled over 50% of the automobile industry.

However, within one decade competition from General Motors reduced Ford's market share to less than 20%. Alfred Sloan, the founder of General Motors, adopted a distinctly different approach. He believed the company should focus solely on core competencies, outsourcing all other processes to expert third parties capable of delivering higher quality components at lower costs. Ultimately, Ford was compelled to transition to outsourcing to remain competitive with General Motors. Nearly seven decades before outsourcing became mainstream, some companies were already implementing such strategies.

In information technology and business process management, outsourcing grew substantially in the late 1960s dues in part to initiatives undertaken by companies. By the 1980s, most large corporations outsourced their IT functions.

The next phase, offshoring, began on a wide scale in the 1990s. Advancements in telecommunications infrastructure enabled companies to relocate some operations to offshore locations offering lower costs and large skilled workforces. China emerged as a primary destination for manufacturing while India specialised in IT and business process outsourcing services. Companies established either captive centres or outsourced to suppliers with operations in these nations.

Outsourcing and offshoring can drive business process innovation within companies. Specific benefits include cost reduction through leveraging external resources at lower prices than internal alternatives; access to hard to find skill sets by operating globally; and quality management through engagement with expert suppliers incorporating extensive quality control and continuous improvement. Additionally, by outsourcing non-core functions companies can concentrate internal efforts on value-added activities like innovation and research & development.

Digital technologies and strategies are intertwined with outsourcing and offshoring. Digital communication and collaboration tools enabled widespread outsourcing and offshoring. Conversely, these practices permitted access to digital innovations, technologies, and talent pools otherwise absent from internal resources.

12.8.2 Unbundling of the Value Chain

A value chain refers to the set of activities a firm in a particular industry performs to deliver a valuable product or service. Reduced telecommunication costs, improved communication quality, and advancements in digital technologies have led to the emergence of a new hyper-outsourcing business model.

Many start-up companies retain internal ownership of marketing activities and intellectual properties (IPs) related to their core technologies and processes. However, they outsource all other elements of their value chain. This allows them to maintain a lean structure and scale rapidly. Ride-sharing companies like Uber and e-commerce retailers like Amazon outsource in this manner. These organisations internally own their IPs, technologies, and processes while leveraging partners for all other aspects of the value chain.

For example, one Indian cosmetics company owns the IPs for their formulations and brands. However, they outsource all other processes. Specifically, the company outsources research and development to a laboratory in the United States and crowdsources a portion of innovation through platforms like Incentive. Manufacturing is handled through outsourced partners in India and Korea. Products are sold on global e-commerce platforms like Amazon and Sephora, while an international marketing firm manages all brand activities. In this way, the company orchestrates the value chain without owning each element. Through this structure, the company remains lean, fosters innovation, and rapidly builds a global brand presence.

While the above example features a new-age company, even traditional firms are unbundling their value chains. In the automobile industry, it is no longer uncommon for competitors to share components like engines or platforms. This trend began with platform-sharing across different models of the same company (Audi and Skoda). However, it has now extended to sharing across competitors as well (Toyota and Suzuki in India). In today's rapidly changing business environment, companies prefer to internally focus on long-term controllable assets like design, branding, and distribution, while collaborating on all other value chain elements.

12.8.3 Product Innovation

Product innovation involves the creation or improvement of a product or service. Incremental innovation can occur through continuous but evolutionary improvements. A clear example is the annual release of smartphones, where each new model features an updated processor, additional memory, or improved display.

Radical or revolutionary innovation involves launching entirely new products or services. The emergence of ridesharing services exemplifies revolutionary innovation. Revolutionary products are rare and carry greater risk, while evolutionary products are less risky and have become standard. Even when differences between smartphone versions are minor, companies release new models as a demonstration of their ongoing innovation to consumers.

Incremental innovations necessitate limited changes targeted towards existing markets. In contrast, innovations relying on new technologies to reach new audiences represent architectural innovation requiring minimal technology changes to address a new market or disruptive innovation developing new technologies aimed at existing segments.

12.8.4 Product development

The New Product Development (NPD) process is a systematic methodology, from ideation to bringing new products to market. All companies, regardless of size and industry sector or whether in the service or product industry, need an NPD process.

NPD is complex and fraught with risk. According to a study, new products are introduced each year, and most of them fail. Therefore, it is essential to have a formal framework and a team that works exclusively on product development.

The NPD framework can be used to improve existing products or develop new ones.

Customer focus:

Focus on the customer?

How are customers evolving?

Is there a gap in customer expectations and product performance?

Do customers regularly expect new products?

Competitive advantage

How competitive is the industry?

How often are products launched?

Who is gaining market share in the industry?

What do they do differently?

Digital strategies and technologies are increasingly disrupting industries and products. It is therefore essential for companies to have a prescribed process to keep them agile and relevant in the market.

There are several frameworks used by companies. At a broad level, however, they all prescribe similar steps.

Market Scan: Company's industry, competitors and customers. Market scanning needs to be a continuous process and many companies outsource this process. Some sources of market information are as follows:

Industry report

Industrial organisations

Market news

Feedback from the sales team

Internal win and loss analysis

Customer feedback (informal surveys)

Salespeople are an important but often neglected source of market intelligence, and many companies have established a formal process for gathering and processing information from their sales force.

A win-loss analysis is an analysis of sales wins and losses versus the competition. It gives the company a clear indicator of customer requirements and why the company is winning or losing in the market.

Initial market analysis: A preliminary test is conducted with a sample of potential customers. Some of the steps taken to test the concept include the following:

- Creating a product storyboard
  - What is the product about?
  - What customer problem does the product solve?
  - How does it differ from other currently available solutions?
  - What benefits can the customer expect (price/features/ease of use/ etc.)?
- Identification of a set of customers/potential customers with whom the story-board can be tested
  - Current customers
  - Friendly customers
  - Potential customers
  - Non-customers (customers currently using alternative product categories)

- Conduct qualitative interviews

  - Get customer feedback on display products.

  - What is the difference between their expectations and these products?

  - Introduce new concepts to customers

  - What is their take on the concept? Positive negative

  - How can the concept be improved?

  - Would they buy the product if it were available?

12.9 Digital Platform Development

In traditional automotive manufacturing, engine and chassis platforms are shared across product lines and companies. In the digital world, platforms like these are changing the way new products are developed and brought to market. The most common example is the Android platform developed by Google, which allows companies to launch new services using the platform. There are many other such platforms that provide the underlying infrastructure and tools to develop and operate customised and customised services and processes. These platforms are now expanding to combine physical products with technology, sometimes referred to as Physical.

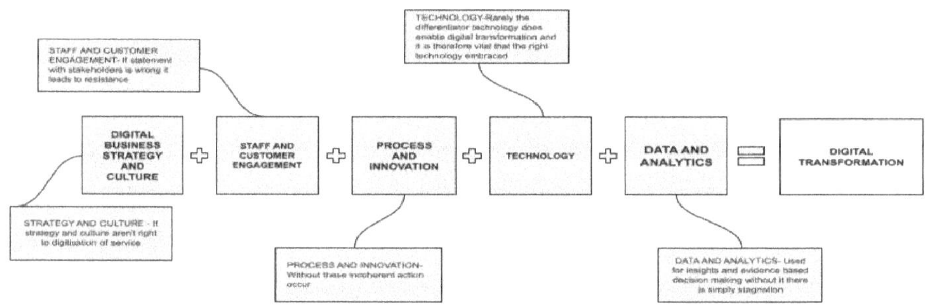

Exhibit 12.4 Digital Transformation Process

12.10 Challenges in Adopting Industry 4.0 Technologies

Many established companies have core products and services that account for a significant portion of their revenue base and are profitable. These established companies sometimes overlook disruptions and disruptors in their early stages, as these disruptions conflict with the core revenue-generating products of the company. A well-documented example of this is how photography giant Kodak was unable to transition to digital despite developing the first digital camera.

Their traditional business model at that point in time (selling camera film) did not allow for development of this concept.

Kodak, a company headquartered in Rochester, New York, was founded in the late 1880s. For over a century, the company dominated the photography industry with very few competitors. The company followed a "razor and blade" business model by selling inexpensive cameras and earning revenue through film sales. It held over an 80% market share in both these product categories in the United States. In the mid-1970s, the company invented the digital camera that did not require film. The company felt that introducing the digital camera would harm its business model and did not actively pursue digital cameras. Within a decade, digital cameras became the standard, and Kodak rapidly lost share to Japanese competitors Canon and Sony. In subsequent decades, cameras became integrated into cellular phones, and the entire camera industry was disrupted. In 2012, one of the most valuable brands in the world filed for bankruptcy protection.

12.11 Role of Leadership and People to implement Digitization in Industries

Transformational and Transactional Leadership

As first proposed by Pulitzer Prize winner James MacGregor Burns in his seminal 1978 book Leadership, transformational leadership is a concept describing how leaders interact with and shape society. Burns classified leadership styles as transactional or transformational. Transactional leaders guide through social exchange, providing incentives and mentoring others to become transformational leaders themselves. American scholar Bernard M. Bass developed Burns' ideas into the renowned Bass transformational leadership theory, also known as the four I theory.

According to this theory, transformational leadership can be explained through four attributes:

- Idealised influence refers to a leader sharing a compelling vision and mission that the team trusts and is motivated to achieve through accepting innovative solutions.

- Inspirational motivation involves a leader clearly and effectively communicating in a way that generates optimism and enthusiasm.

- Intellectual stimulation encourages the team to explore new problem-solving approaches in innovative, breakthrough ways.

- Individualised consideration means the leader provides personal attention and development to each team member so all feel valued and important.

Digital strategies are high priorities for corporate boards and chief executive officers (CEOs). According to a 2018 Gartner study of CEOs, growth remains the top priority for most, though what drives growth has changed— digital strategies are now the key driver.

The survey found that every executive committee member must be engaged in developing, implementing, and managing digital strategies. CEOs are concerned that many executives lack sufficient digital skills to help the company's future. Executives in information technology, sales, manufacturing, technology, supply chain, and human resources were cited as needing the most development of digital skills. CEOs stated that implementing a company-wide digital strategy would be challenging without these skills being developed across the entire executive team.

When asked which organisational competencies companies need to develop most, CEOs named talent management, followed closely by technology enablement, digitalization, and data management. CEOs emphasised that data literacy and a data-centric approach are critical to overall company growth.

Leading a successful digital transformation requires various roles and responsibilities from CEOs. Dan Shapero, vice president of Global Solutions at LinkedIn, well summarised the key attributes: do they build great teams, understand technology's business implications, adapt to rapid change, operate at both high and granular levels, and build trust across the organisation?

Some characteristics of a successful digital CEO include:

Vision - A digital strategy combines elements of innovation. Without clear short- and long-term goals, transformation is impossible.

Collaboration - Successful digital transformation requires collaborating internally across functions and externally with partners. CEOs must implement knowledge-sharing processes.

Experimentation - Digital strategies require encouraging experimentation given ambiguity. Experimentation must be data-driven with contingency planning and rapid modification.

Team Building - A digital strategy relies on strong execution teams, starting with a chief digital officer. CEOs ensure an embrace of change across the organisation.

Executive Focus - CEOs provide governance, milestones, feedback, and properly aligned incentives to execute the new vision, such as for both legacy and digital IT teams in a transforming company.

12.12 Role of Investors and Boards in Digital Transformation

The board plays an important role throughout the digital transformation process. They are involved in encouraging companies to begin their digital journey, approving necessary financial resources, monitoring impact, providing strategic direction, helping build networks, and making course corrections when needed.

Boards are responsible for balancing a company's short-term goals with creating sustainable long-term advantages. As the BDO Cyber Governance Survey found, public company board directors are increasingly recognizing both the necessity and competitive benefits of embracing a digital transformation strategy. Boards must understand the benefits and risks of digital transformation, as well as the downsides of failing to go digital.

Many digital companies, especially start-ups, focus on growing their user base in the short and medium term. This growth focus comes at the cost of profitability and requires significant financial resources to sustain high burn rates. Companies need a constructive relationship with their boards during this high-risk growth phase.

Board members typically oversee the digital transformation of multiple companies across various industries. They provide strategic insights and best practices from a range of sectors to company leadership. In this age of disruption, leaders should consider the perspectives of non-competitors. Investors and boards can serve as valuable sources of such information.

12.13 Chief Digital Officer

Companies around the world are creating the new role of CDO or Chief Digital Officer to drive digital strategies and transformation in their business. Born digital companies have a digital need to create this culture. However, traditional companies must create this culture under the leadership of the CEO and board of directors. Digital strategies involve all parts of the organisation, not just technology. In traditional companies, these roles are divided between Chief Marketing Officer (CMO), Chief Business Officer, Chief Information Officer (CIO), Chief Technology Officer (CTO), Chief Financial Officer (CFO), Chief Human Resources Officer (CHRO), and Chief Executive Officer. Operations Officer, among others. The single point of convergence of all these functions is the CEO. As the CEO has multiple priorities, it was necessary to create a new foothold for this activity. This has led to an increase in the position of CDOs in companies around the world. A 2017 study by Strategy found that around 19% of the world's leading companies have created a CDO role, mostly in the past five years.

There is considerable confusion between the roles of CIO and CTO as both focus on technology. CIOs are generally focused within a company and are responsible for IT applications and infrastructure. The CIO deals internally with business groups, the IT team, and externally with vendors of IT services and products. A CTO, on the other hand, is externally focused on customers and works to develop products and services that are sold to customers. For example, an IT products company might have a CIO responsible for all internal infrastructure, while a CTO focuses on designing and developing products sold to customers. The CDO is responsible for developing and implementing digital strategies and does not need to have a technology background. In other words, the CIO focuses on continuity, the CTO focuses on growth, and the CDO focuses on change.

Each of these roles requires different skills, and companies have been found to struggle when the CIO or CTO is given CDO responsibility. According to a global survey of IT leaders, companies became effective in digital transformation only after creating the role of a specialised CDO. The study found that half of organisations with a dedicated CDO had a clear digital strategy, compared to only 21 percent without a CDO.

12.14 Profile structure and reporting

Another issue that many companies have struggled with is the ideal profile of a CDO and who the CDO should report to. It has been argued that CDOs need to bring together several functions in a company, and should be dedicated or permanent employees. On the other hand, permanent employees may not be the best fit for change management. An outsider who can bring new ideas and best practices from other industries and companies can be a better agent of change. Many companies have brought back experienced professionals who had tenure and then gone to work for other companies to become CDOs. Such individuals meet both the tenure conditions and the range of experience. Diversified conglomerates have also chosen individuals who have worked in multiple group companies, as this provides them with a wide range of experience.

In the early days of this role, CDOs reported to CIOs. Companies soon realised that this was not ideal, and today most CDOs report directly to the CEO.

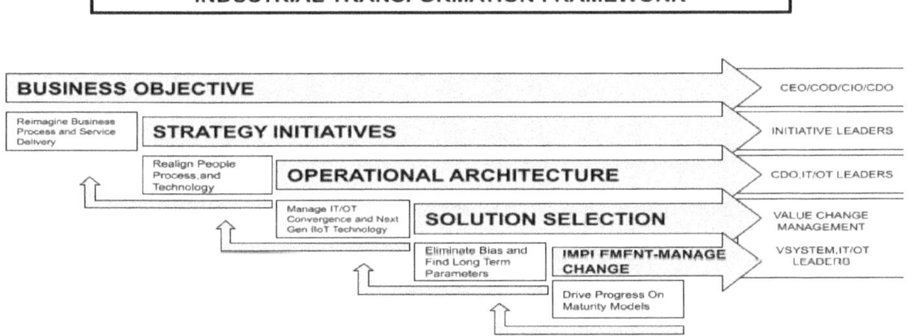

Exhibit 12.5 Industrial Transformation Key Role Players.

Key responsibilities

The CDO has several responsibilities related to the design and implementation of digital strategies.

Designing the digital strategy: The CDO's primary role is to formulate the organisation's digital strategy. In most companies, CDOs invite several external consultants to provide their perspectives on potential digital strategies. Based on this input, the CDO works closely with the CEO to formulate the company's digital strategy.

Prioritise elements of digital strategy: A transforming company will undergo massive changes across functions. This takes time and resources, so prioritising the elements of transformation is essential.

Building an innovation ecosystem: Implementing digital technologies required a wide range of skills including consulting, design, implementation and testing. For large transformations, these skills are not available in-house or entirely with an external partner. The CDO needs to build an ecosystem of partners, both internal and external. In many cases, this also includes startups, research labs and academic institutions.

Change management: Digital transformation involves changes across the organisation, including changes in people, culture, working methods and technology. The CDO is responsible for working with various stakeholders to build a culture of change.

Implementation Management: The CDO works with cross-functional teams to ensure implementation of the transformation. This includes program and project management, administration and reporting.

Measuring ROI and Objectives: The primary reason for digital innovation is to create value and stay competitive. The CDO sets up a mechanism for measuring and reporting the progress of the transformation, including the achievement of return on investment.

Digital Talent Development: The CDO works closely with HR to attract and retain top talent and build digital capabilities.

Communication: The CDO is also responsible for communicating the goals and status of the transformation to both internal and external stakeholders.

The CDO role is gaining recognition across companies. This is reflected in the number of CDOs who have taken over as CEOs of companies.

Roles and Impact on Teams: The Future of Work

The role and impact of digital transformation on employees has often been referred to as the 'future of work'.

Companies need to bring together many different capabilities to successfully implement digital strategies. On the front-end, they are strategy experts, design thinkers, business modellers and business analysts. On the digital implementation end, they require solution architects who can design the appropriate technology architecture. These architectures must be supported by a range of digital experts, including blockchain, IoT or AI specialists. Companies also need expertise in change management and program and project management.

Digital transformation affects all aspects of people management including talent acquisition, talent management or talent engagement.

12.15 Talent acquisition

Talent acquisition begins with identifying the type of HR required in the company, then developing recruiting resources and moving on to the actual hiring and on boarding process.

To build a successful digital organisation, companies need to attract and hire a new skill set. A report released by the World Economic Forum highlights several technology trends that will be essential for organisations to transition to digital broadcasting.

Because many of these skills are new, in short supply, and may not be in demand for a long time, companies will need to innovate their talent acquisition strategies. This could include moving to hybrid models where some employees are full-time while others, especially those with specialised skills, can be drawn from the talent ecosystem.

In addition to in-house and external employees, companies can use contract workers, gig workers, and latent crowdsourcing. Contract workers are provided by third-party staffing agencies and are typically used for short-term, high-volume work. Gig workers are independent contractors who have professional skills and do not want to be tied to any organisation. While gig work is increasingly popular among millennials, it is struggling to recruit organisations and workers as a class of workers. Platforms like Flex Jobs connect gig workers with employers.

The platforms have also led to an increase in crowdsourcing of workers. For example, companies post problems that require complex analysis on platforms like Kaggle. Teams and individuals from around the world compete to solve these problems for cash prizes.

The entire talent acquisition process from talent acquisition, shortlisting, interviewing, accepting offers and on boarding can also be completed on digital platforms.

Future of Jobs

Design and adaptive organisation

An organisation striving to be a digital leader must anticipate and adapt to change. As CEOs and CDOs develop their digital strategies, they need to assess whether they have the right teams for change. This includes people in key leadership roles who must be able to plan and deploy change. Management also needs to gauge their teams and assess whether they can move existing people into new roles or whether they need to hire new people with the required skills.

In most transformations, companies deploy their newer digital strategies while the existing legacy business continues to operate. Organisations must decide what proportion of existing resources to commit to the change process without disrupting existing business. People strategy must include evaluating the processes and systems that need to be adapted to align with the new business strategy. Corporate culture is another critical element of human strategy.

A company must determine what aspects of its culture must evolve and whether informational, functional, or business forces are likely to be obstacles to how the company evolves.

Ability of all internal staff and departments to effectively collaborate, collaborate and share information.

The ability of all employees to connect with external stakeholders including customers, suppliers, industry experts and consultants, even competitors.

The ability of all employees across the organisation to innovate and experiment without fear of failure, supported by the right incentives and a culture that supports innovation.

To enable this culture of change and innovation, a company needs to empower and support individuals and teams through several initiatives.

Building self-driven teams: All bottom-up transformation starts with empowering self-motivated teams. Overburdening teams with structures and processes reduces collaboration and limits innovation. Smaller groups are better at developing innovative solutions.

Breaking down traditional barriers: Most large organisations operate in silos with limited flow of information and ideas across those silos. The exchange of ideas across these forces will help to gain a comprehensive view of the problems and develop effective solutions.

Creating spaces: Organisations need to create space and time for innovation. These could be physical spaces such as innovation labs or virtual spaces where "innovation teams" are formed to discuss and come up with ideas for the future. These spaces allow employees to think "out of the box" and provide a non-judgmental environment for the free generation and exchange of ideas.

Industry 4.0 presents opportunities for transformation, but it is not just about the development and adoption of SMART technology. And these are not evolutionary or transformational changes. Industry 4.0 is a different way of thinking that will allow us to jump into a different future. In order to reap the transformative benefits of Industry 4.0, we need to embrace cutting-edge technology in the best way to achieve better results from a wider range of interests. This means rejecting traditional and outdated paradigms, such as a too narrow focus on efficiency without considering resilience and the ability to function in times of crisis and disruption to wider distribution and production systems. We must question everything that came before.

For example, traditional manufacturing is based on production runs and batches. Central to Industry 4.0, however, is the opportunity to "produce a single batch through a highly flexible and intelligent manufacturing process," enabling a production line that communicates and adapts in real time to mass produce a unique product. This change represents a more holistic assessment of "effectiveness and efficiency", moving away from the centralised approach of mass production to a more nimble and distributed concept. This concept is almost essential to succeed in the crumbling global economy, which is already evident. It is likely to be more visible in more sectors and functions, given the impact of the pandemic, an assertive China, technological change and now the return of war in Europe and the rise of a functioning strategic partnership between Moscow and Beijing.

Better use of natural resources

Industry 4.0 often focuses on exploiting the productivity benefits of reduced resource sources. Such opportunities are real, but they are only part of the picture. A great opportunity lies in the re-evaluation of outdated production processes. Changing our thinking will allow us to modernise the economy and business environment, making non-profits viable one day. The shift away from 20th century lean process management focused on improving inefficient processes and reducing process waste will change the way new ideas are brought to life.

Key Notes:

Topic	Keynote
Meeting Customer Expectations	Meeting customer expectations in the digital age
The Role of Governing Entities	The crucial role of governing entities in digital transformation
Leveraging Third-Party Networks	Utilizing external platforms for sales, marketing, and procurement
Dynamic Leadership and Digital Strategy	The importance of dynamic leadership in digital strategy
Growth and Customer Connectivity	Strategies for growth and enhancing customer connectivity
Turbocharge Process Efficiency	Optimizing process efficiency through digitalization

Go Global and Explore	Expanding global presence through digital initiatives
Attracting Top Talent	Techniques for attracting and retaining top talent
Level of Adoption	Understanding the varying levels of digital adoption
Does Every Company Need a Digital Strategy?	Assessing the necessity of a digital strategy for every company
Process Reconstruction	Adapting to new paradigms in process management
Outsourcing and Offshoring Strategies	Exploring strategies for outsourcing and offshoring

Summary:

Customer Expectations: Meeting customer expectations in the digital age is crucial for business success, as consumers increasingly demand seamless digital experiences across all interactions.

Governing Entities: Governing entities such as private equity firms and venture capital play a significant role in driving digital transformation by providing financial resources, strategic guidance, and monitoring impact.

Third-Party Networks: Leveraging external platforms for sales, marketing, and procurement allows companies to expand their reach and access new markets efficiently. Dynamic Leadership:

Dynamic leadership is essential for successful digital strategy implementation, as leaders must adapt to changing market dynamics and technological advancements.

Process Efficiency: Digitalization enables companies to turbocharge process efficiency by automating repetitive tasks, streamlining workflows, and optimizing resource allocation.

Global Expansion: Digital initiatives facilitate global expansion by providing access to new markets, customers, and partners, enabling companies to scale their operations effectively.

Talent Acquisition: Attracting and retaining top talent is critical for digital transformation, as skilled professionals are needed to drive innovation, execute digital strategies, and navigate complex technological landscapes. Product Innovation: Digital technologies enable companies to drive product innovation by developing new offerings, improving existing products, and delivering value-added services to customers.

Challenges in Adoption: Adopting Industry 4.0 technologies poses challenges such as resistance to change, legacy systems integration, and cybersecurity concerns, which must be overcome to realize the benefits of digital transformation.

Role of Leadership and People: Leadership plays a crucial role in driving digitization efforts, as leaders set the vision, provide direction, and create a culture of innovation and continuous improvement. Additionally, investing in talent development and empowering employees is essential for successful digital transformation.

Trivia Challenge: Industry 4.0

Welcome to the Industry 4.0 Trivia Challenge! Get ready to test your knowledge of the wacky world of Industry 4.0 with these hilarious multiple-choice questions. Each question is paired with some side-splitting facts and figures about Industry 4.0. Let us dive in and have a giggle!

1) Why did the digital strategy cross the road?

A) To meet customer expectations on the other side.

B) To govern the entities on the opposite sidewalk.

C) To leverage third-party networks across town.

D) To avoid the challenges of Industry 4.0 integration.

2) How does a digital strategy navigate the complexity of Industry 4.0?

A) By automating production and reconstructing processes.

B) By relying on leadership to lead the way.

C) By integrating digital platforms into every aspect of operations.

D) By avoiding challenges and sticking to traditional methods.

3) What's a digital strategist's favourite tool?

A) A hammer for smashing through old processes.

B) A magnifying glass for examining customer expectations.

C) A compass for navigating the digital landscape.

D) A crystal ball for predicting future technological trends.

4) Why did the company hire a Chief Digital Officer?

    A) To report on the success of digital transformation.

    B) To lead the charge in adopting Industry 4.0 technologies.

    C) To acquire talent for integrating digital platforms.

    D) To challenge the board on digital strategy decisions.

5) How does a digital strategy deal with talent acquisition challenges?

    A) By outsourcing talent to third-party networks.

    B) By relying on investors to fund talent acquisition.

    C) By offering competitive salaries and benefits.

    D) By avoiding talent acquisition altogether.

6) What's the digital strategist's secret weapon?

    A) A magic wand for instant digital transformation.

    B) A roadmap for navigating the digital landscape.

    C) A crystal ball for predicting future digital trends.

    D) A shield for protecting against digital threats.

7) How does a digital strategy handle resistance to change?

    A) By ignoring resistance and forging ahead.

    B) By engaging leadership to address concerns.

    C) By avoiding digital transformation altogether.

    D) By waiting for investors to approve change initiatives.

8) Why did the digital platform go on vacation?

    A) To recharge its digital batteries.

    B) To avoid the challenges of Industry 4.0.

    C) To explore new talent acquisition strategies.

    D) To report back to the board on its performance.

9) What's the digital strategist's favourite board game?

    A) Monopoly, for mastering the art of acquisition.

    B) Chess, for strategic planning and manoeuvring.

    C) Scrabble, for spelling out digital transformation.

    D) Clue, for solving the mystery of Industry 4.0 challenges.

10) How does a digital strategy measure success?

A) By meeting and exceeding customer expectations.

B) By reporting back to the board on its achievements.

C) By integrating third-party networks into its operations.

D) By avoiding challenges and maintaining the status quo.

Abbreviations:

Abbreviations	Terms
CE	Customer Expectations
GE	Governing Entities
TPN	Third-Party Networks
DL	Dynamic Leadership
PE	Process Efficiency
GE	Global Expansion
TA	Talent Acquisition
PI	Product Innovation
CA	Challenges in Adoption
LP	Role of Leadership and People

Chapter 13

Transformation from Industry 3.0 to Industry 4.0

Learning Objective:

Upon completing this section, readers will have comprehensively understood the

- Transformation journey from Industry 3.0 to Industry 4.0.

- Understanding the evolution and innovation driving this transition.

- Different strategies for implementing Industry 4.0 and the stages in its evolution.

- Industry 4.0 strategy and its impact on shaping global trends.

- The role of LEAN Production Systems.

13.1 Prepare for Industry 4.0

Advances in technology have brought people, businesses, countries and the entire business world closer together than ever before, and these trends will continue to increase.

13.1.1 Preparation

Since production is affected by concepts such as the internet of things, cyber-physical systems and cloud-based production, it is necessary to evaluate whether companies are ready for Industry 4.0. Industry 4.0 holds great potential for technology users in every field. However, there are still many unresolved questions, uncertainties and problems. The Work Plan is designed to meet this need and provide insight. It also highlights the significant challenges many companies still need to overcome on their way to Industry 4.0 readiness.

The basic point in this understanding is that smart factories and smart products represent dimensions related to the physical world, while smart operations and data-oriented services represent physical dimensions.

According to this concept, Industry 4.0 can be called the combination of the physical world and the virtual world. Details of the planning model are explained below

13.1.2 Strategy and Planning

Business 4.0, in addition to improving the existing process by using digital technology, also offers new opportunities to create completely new business models. Current openness and leadership can be tested using the following models:

- Business 4.0 Implementation of existing information strategies
- Control strategies for better performance through measurement
- Industry 4.0 Business-related measurement
- Understand technology and management innovation
- Understand the current state of R&D.

13.2 From Industry 3.0 to Industry 4.0

The evolution of technology business can be traced to various stages, each marked by the advancement of technology. technology and change. Understanding the historical context of Industry 3.0 and the subsequent emergence of Industry 4.0 provides a better understanding of the rise path of the Industrial Internet of Things (IIoT).

13.2.1 Industrial Computerisation

Industry 3.0, often referred to as the age of automation, envisages the integration of computers into business processes. This phase began at the end of the 20th century and marked a major shift between manual labour and the introduction of machine control. The advent of programmable logic controllers (PLC) and computer numerical control (CNC) machines has revolutionised manufacturing by enabling precise and automatic control of machines.

These tools facilitate productivity and consistency by automating many business processes.

CNC Machine Tools: Computerised machining tools bring new levels of precision to the manufacturing process. These machines, controlled by computer programs, enable automation and precise shaping of materials, laying the foundation for progress in sectors such as aviation and automotive.

13.2.2 Limitations of Industry 3.0

Although Industry 3.0 represents a significant leap forward in technology and automation, it also has its own limitations. These systems are often isolated and cannot communicate with each other. The fact that the data flow is limited to a single machine or slide control system prevents the entire production process from being seen. This limitation has created the need for a more connected and smarter business environment.

Moving to the Fourth Amendment

New Information on Intelligence, Information and Management Intelligence Industry 4.0 paves the way for a new business model based on new wisdom.

In one corner, there are e-commerce models such as Google, Facebook, Tencent, Airbnb, Uber, Spotify... They are very effective in stock market sharing with their annual growth rate. For both figures (although some of the results are questionable), there is no repeat of history and they are well known for their cannibalism in oligopolistic businesses.

In the other corner are the ever and undeniably successful companies such as Rockwell, ABB, Schneider, Yaskawa, Mitsubishi, Siemens with a proven track record in sustainability. They have deep knowledge, a long-term perspective and are backed by significant investments. This allows them to compete based on clear regulations (and financial criteria) in a less dynamic market, while ensuring their commitment to creating shareholder value and meeting the need for other partners to help build equity.

This defence of "culture" makes no secret of the need to take the foundations of the Fourth Revolution (knowledge management, the Internet of Things and other technologies) to pave the way for innovation. Businesses are built on new skills.

However, companies that want to achieve a successful digital transformation need to know three basic management principles in order to keep up with the new e-commerce model

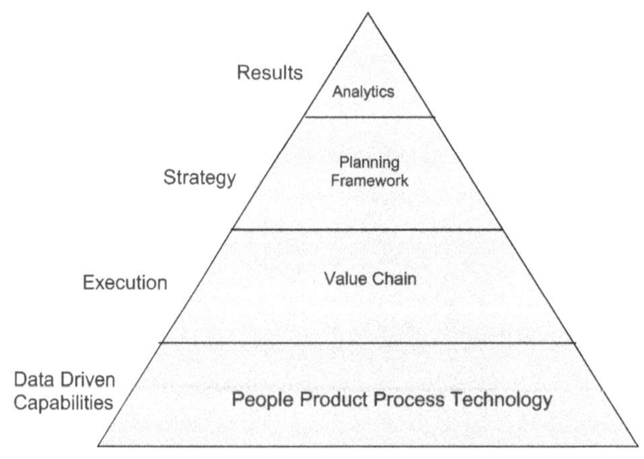

Exhibit 13.1 Industry 4.0 work segmentation

Ability to engage with customers, keep customers into the "factory" and allow them to participate in creating products and services customised to meet their needs. This means changing the processes of the organisation and forcing decision-making that in the future must be more disciplined and focused on business processes, and people refuse to do this.

The company's interaction with its customers creates information that allows it to predict the customer's behaviour. This data is tracked faster than the sales forecast.

Demand management is a complement to, and sometimes a substitute for, the primary use of competitive strategy. This new policy of good management makes it possible to manage uncertainty and compete with demand even in stressful situations.

Ability to adapt through quick and flexible responses, business connection and turnaround proximity. The distance between today's manufacturers and their products must be overcome by developing business ecosystems equipped with approved information and communication systems and new rules.

Customised solutions should be more important than reducing cost and the ability to create dynamic relationships rather than long-term relationships. The contract is more important.

The relationship between customers is to create value based on the economic model, advantageous with powerful people, capital nature with other crowded social models, and offer new solutions to customers.

13.3 Innovation

Regarding these skills, we need to create the business:

- Innovation in the organisation of information and processes

- Innovation in supplies and methods, assistance

- Innovation in the new business model

The first two elements take priority. Some elements are inherent. The ability of "factories" to learn and interact with customers is an innovation in itself, influenced by the design and management of processes, production, and decision-making.

Whereas innovation can be seen or analysed by a SMART Goal and Business objectives, where Vision comes to action, which makes it possible to profit from new solutions rather than changes and agreements between members.

The main problem of the business sector is the result of the new business model

However, the great competition in the business world is the result of the new copy. Technological development makes it possible to include new services in products, thus increasing or changing the final price, in other words the services of the products.

In addition, the digitalization of the economy has new consumption and business models that go beyond the transfer of ownership, focusing on pay-as-you-go (e.g. rental, subscription and payment).

Innovative industries are expanding their action work and finding great roots in the fields of renewable energy, smart transportation, circular economy, optimization of natural resources and management of smart cities and buildings.

So, if companies get the management system quickly and efficiently, the new e-commerce model is very good, they will not only achieve returns but also achieve business interests.

In addition to these management skills, companies also need to strengthen their ability to explore new partnerships. Create shared benefits and prioritise sustainable medium- and long-term personal "win-wins."

Only in this way can businesses respond, as in the past, to the big challenges facing society today: climate change, depletion of natural resources and the elderly.

Doing Business in Industry 4.0 in established organisations with a long history is more difficult because it has many consequences of Development. In this section, we will explain business transformation and how business has changed over the last three hundred years.

Business transformation means changing the company's business. Change is driven by other factors such as customer preferences, competition and the regulatory environment. It can also result from internal factors, such as changes in culture or the availability of new processes or technologies. It can be incremental or revolutionary.

Progressive transformation is incremental and occurs in incremental steps. It continues for a period of time and allows the business to adapt to changes. Transformational change can occur through technology projects or process improvements through Six Sigma or kaizen projects. It is usually not organisation-wide and is limited to changes in functions. Revolutionary transformation is the implementation of major changes in a short time. This change can result from the use of new technologies or business models and can affect all aspects of the business.

Changes at the operational level

Changes in the action Working, making significant changes to the processes and the way the company works. For example, it may include issues such as creating new systems for data entry or replacing key systems.

Employment changes have immediate and long-term effects. Manufacturing companies often experience many changes in their operations as new production and automation methods are developed.

Changes at the Business Model Level

Business models define how a company creates and captures value. In order to change the means of creating value and capturing costs, companies need to plan their internal and external processes differently. The consequences of changing a business model generally carry higher risk and are less predictable than those of changing operations. There are several recent examples of this type of change. One of the most talked about articles about changing the business model is how Amazon is constantly changing the way books are produced and sold with its e-commerce model.

Transformation at the Strategic Level

Strategic transformation means changing the essence of the organisation. This is a combination of business and business model changes. This is the riskiest change and if it fails it can be disastrous for the organisation. If this change is successful, the company will be far ahead of its competitors. Amazon and Apple are two examples of companies that have initiated and successfully implemented revolutionary changes.

13.4 Different Strategies to Implementation of Industry 4.0

Evolution of Industry 4.0

Industry 4.0 is the way for the manufacturing industry to find future business development. The rise of machines and robots has been driven by computers, information technology, and semi-intelligent robots that have replaced many workers. Businesses will scale quickly and achieve results, keeping in mind the needs for ease of business and value-added level of operations.

Industry 4.0 is the most important thing for all developers. No leader wants to ignore this in his strategy. Leaders across industries are eager to understand

Industry 4.0 and its implications for transforming organisations by leveraging data, creating new insights, and changing business models.

But industry reports and analysts offer a mixed view of the practice. While forward-thinking companies are seeing results in implementing Industry 4.0, many face uncertainty when deciding how to implement and scale the business.

Traditional project management methods with fragmented business processes are not sufficient to meet the needs of Industry 4.0. Industry 4.0 applications must have a deep understanding of business needs and opportunities, Industry 4.0 concepts and technologies, and how they relate to each other.

Advances in technology could lead to the next economic revolution. Looking to the future, the integration of physical and virtual worlds into cyber-physical systems will have a significant impact on all aspects of production. Compared to Industry 3.0, traditional businesses expect an unprecedented combination of information, communication and production, including:

Smart sensors communicating with each other through the Internet of Things allow real-time monitoring of the production process and collecting data rights of the production process.

Big data about people, machines and factories.

Meet a new form of human-computer interaction. This includes the development of augmented reality technology that leverages smart devices and mobile devices.

The increase in cloud computing, data storage, cloud storage, computing power and network connectivity has previously made data analysis work impossible. This enables data hosting and ensures that data can be accessed from anywhere at any time.

Advances in analytical capabilities are needed to improve product quality. Big data analytics enables seamless processing of big data. Many analyses are needed to improve the performance of the business.

13.5 Stages in the evolution of Industry 4.0

Visibility

The first level is the visibility of the result on the shop floor, which is important for Industry 4.0 to work without a closed loop for feedback in the workplace. The workshop is now isolated and unable to communicate with other departments of the company, but is integrated with other departments through information flow.

In terms of Industry 4.0, this level provides communication protocols for closed-loop connectivity, interoperability of systems and devices from various vendors, and seamless interrelationship. Using sensors and Internet of Things (IoT) connectivity, the importance of production systems can be transferred to tools such as dashboards and charts for monitoring purposes.

A major food and beverage brand generated real-time production data from factory equipment to increase factory visibility. The initiative increased production lines and factory efficiency, and made the factory more profitable. The change also enabled the elimination of most manual labour, including printed materials.

Analysis: Based on an understanding of what happened, the next question is why? Answering these questions helps organisations understand the root cause of the problem.

In the oil and gas industry that produces slurry, moving parts of rotating machines that need to operate continuously experience problems and frequently fail. This should predict the machine's performance and prevent downtime. Testing equipment is classified by analysing before and after the crime problem. Such analysis helps identify three sets of data that can be used to develop predictive models to verify failure.

Simulation: Next come the "what" and "why" questions, followed by the "how" and "what is" questions. Simulations can answer questions about how the system will behave under different conditions. Predictable fault signals in the virtual machine can be detected long before the physical machine fails, allowing for timely action.

In digital twin projects in the aviation industry, virtual models of aircraft landing gear are designed to detect failure modes or ways in which the system may not function properly. Thirty-four sensors were placed on the surface to measure parameters such as pressure and temperature, providing information for early detection. Data is collected every second during landing and takeoff and placed in a virtual model. This model is used to diagnose existing problems and estimate the remaining service life of the landing gear. The system ensures that flights are on time and that the landing gear functions well throughout its lifespan.

13.6 How to implement an Industry 4.0 strategy – Implementing method

The term Industry 4.0 involves the adoption of new technologies and the creation of new forms of expertise to increase efficiency. There are different suggestions for businesses to follow the Industry 4.0 strategy and increase their chances of success in the digital environment.

One of the principles of Industry 4.0 is to interconnect machines, systems and cells to create intelligent networks based on value chains that can operate independently and tightly manage each other.

When implementing Industry 4.0, we must pay attention to the following differences and expectations:

Integration: All processes must be seamless, interactive and collaborative. The production process does not only rely on the previous process. It involves more than people, machines and processes. The entire environment is flexible, connected, and shares valuable information. Using technologies such as the Internet of Things (IoT), everything can be connected, communicated and interoperated. Multiple SMART technologies are required to ensure seamless integration.

Digital Twin (Simulation): Create a simulation or virtual model of a physical system or process from sensor data or other SMART devices to help create models for testing, research and development (PoC). This does not affect the actual process. This will reduce time and increase revenue.

Autonomy: Industry 4.0 enables different machines in the smart factory to make decisions independently and independently. It does not affect the growth of the organisation. This can be done through a data management system.

Flexibility: Flexibility is also very important when it comes to Industry 4.0. Smart factories can easily adapt to changes. By designing and building products, production systems and even modular conveyor belts can easily adapt themselves to the products to be made. In a smart factory, production lines can be quickly implemented, expanded or replaced because processes do not require major changes.

Services on the Internet: Industry 4.0 enables the integration of physical capability into digital devices and physical processes to create smart factories. The Internet of Things (IoT) is a rapidly growing technology contributing to the success of Industry 4.0. The Internet of Things has created business opportunities for many businesses.

13.7 The Fourth Industrial Revolution: Shaping Global Trends

Globalisation refers to the increasing interconnectedness of economies and cultures worldwide. Advancements in technology, transportation, and communication drive globalisation. It has profoundly impacted the global economy by creating new opportunities for both businesses and consumers. However, it has also presented many challenges such as inequality, environmental degradation, and security risks.

Industry 4.0 refers to the ongoing automation of traditional manufacturing and industrial practices through innovative technologies. The convergence of the physical, digital, and biological spheres characterises Industry 4.0. Emerging technologies like the Internet of Things, artificial intelligence, and big data analytics underpin it.

LEAN manufacturing aims to minimise waste within a production system through a systematic approach. It focuses on identifying and eliminating non-value-added activities between receiving raw materials and delivering finished goods to customers. LEAN manufacturing has the potential to increase productivity, save expenses, and improve quality for enterprises.

Smart, connected businesses can collect, analyse, and share data in real-time to facilitate better decisions. They leverage technology to connect with customers, suppliers, and partners in novel ways. Such businesses exhibit greater agility and adaptability to better satisfy customer needs in a rapidly changing environment.

Smart factories use digital technologies to automate and optimise manufacturing processes. Sensors collect production data from the supply chain to the assembly line. Artificial intelligence and big data analytics then analyse this data to identify areas for improvement. Smart factories can help businesses produce products more efficiently and effectively while reducing waste and costs.

How are these concepts interrelated?

Globalisation, the Fourth Industrial Revolution, LEAN production systems, smart and connected business perspectives, and smart factories are all interrelated. Globalisation has created new opportunities for businesses to expand their reach and operations to new markets. The Fourth Industrial Revolution enabled companies to automate and optimise their production processes, making them more efficient and productive.

LEAN production systems help businesses identify and eliminate waste, improving efficiency and reducing costs. Smart and connected business perspectives enable businesses to collect, analyse, and share data in real-time, making better decisions and meeting customers' needs in new ways. Smart factories culminate all these trends, using digital technologies to create the most efficient and productive manufacturing facilities possible.

Interrelation of Key Concepts

There are linkages between several important business concepts. Globalisation has provided companies opportunities to expand into new markets internationally. The Fourth Industrial Revolution allows organisations to automate and optimise production processes, increasing efficiency and productivity.

LEAN manufacturing systems help identify and remove waste from operations, boosting efficiency and lowering costs. Smart, connected ways of working enable data collection, analysis, and sharing in real-time, supporting improved decision-making and new approaches for satisfying customers. Smart factories bring together these trends, employing digital technologies to establish manufacturing facilities that are highly efficient and productive.

Economic Interconnectedness

International Trade: Globalisation is characterised by heightened intercontinental exchange of goods, services, and capital. Nations have established intricate economic relationships through trade agreements, creating complex webs of interconnectivity.

Financial Integration: Events in one region can have global economic effects due to integrated financial markets where stock markets, currency exchange, and investments are intertwined. This shapes the economic landscape on a worldwide scale.

Technological Advancement and Communication

Digital Revolution: Technological advancement, especially the digital revolution, has accelerated globalisation by reducing communication barriers through the internet, mobile connectivity, and digital platforms that enable real-time interactions across borders.

Global Collaboration: Technology facilitates seamless cross-border collaboration, allowing businesses and individuals to work together virtually through platforms for meetings, shared workspaces, and instant exchanges—integral components of international operations.

Emerging Global Challenges

Environmental Sustainability

Climate Change Concerns: Human activities intensified environmental impacts from globalisation necessitate collaborative solutions to challenges such as climate change, pollution, and resource depletion that ignore territorial boundaries.

International Agreements: The need for worldwide coordination is evident in initiatives like the Paris Agreement, where nations unite to jointly address climate change through collective goals. Environmental sustainability demands harmonised global action.

Supply Chain Resilience

Complex Supply Networks: Intricate, sometimes fragile supply chains have resulted from an interconnected global economy vulnerable to disruptions that reverberate across production and distribution systems.

Pandemic Preparedness: COVID-19 highlighted the importance of resilient supply chains, as border closures, transportation disruptions, and shortfalls emphasised the need for increased flexibility and adaptability to enhance network endurance against systemic shocks.

Globalisation shapes the modern world through economic interdependence and technological progress. However, challenges like environmental protection and supply chain stability necessitate collaborative international solutions due to their planetary scope. As global crises emerge, coordinated transnational cooperation will be paramount to building a sustainable and robust future for all.

13.8 LEAN Production Systems

Origin and development

Toyota Production System (TPS): The foundation of LEAN production can be traced back to the Toyota Production System, which Toyota developed in the mid-20th century. The goal of TPS was to eliminate waste, improve efficiency and increase overall productivity.

Emerging from the challenges of post-World War II Japan, TPS pioneered concepts such as just-in-time manufacturing, where resources are used exactly when needed, and continuous improvement with an emphasis on continuous improvement processes.

The principles of LEAN

Value Stream Mapping: LEAN emphasises understanding and mapping the entire value stream, identifying areas of waste and streamlining processes to create value for the customer.

Just-in-Time (JIT): JIT manufacturing minimises inventory by producing items only as they are needed in the production process. This reduces transportation costs and ensures a more flexible production system.

Kaizen (continuous improvement): Kaizen involves a culture of continuous improvement where employees at all levels are encouraged to identify and implement incremental improvements in their work processes.

Respect for people: LEAN places great emphasis on respecting and involving employees in decision-making processes, recognizing their expertise and contribution.

Integration with Industry 4.0

Synergies and Complementarity

Common Objectives of Efficiency

Both LEAN production systems and Industry 4.0 share the common goal of improving efficiency and responsiveness in manufacturing processes. LEAN principles align with the objectives of reducing waste and optimising resource utilisation.

Data-Driven Decision Making

Industry 4.0 technologies, such as IoT, AI, and big data analytics, complement LEAN by providing real-time data and insights. This data-driven approach enhances precise decision-making, allowing for more informed and efficient operations.

Integrating LEAN in Modern Manufacturing

Industry 4.0 introduces smart manufacturing concepts where machines, systems, and processes are interconnected and digitally optimised. LEAN principles can be integrated into modern manufacturing by leveraging technology for advanced process optimization. Predictive maintenance capabilities of Industry 4.0 align with LEAN principles by preventing downtime and ensuring equipment is functioning optimally. Digital technologies enable quick adjustments to production schedules, responding to changes in demand or unforeseen disruptions.

The integration of LEAN production systems with Industry 4.0 represents a synergy of traditional efficiency-focused principles with cutting-edge digital technologies. By leveraging the strengths of both, organisations can create a manufacturing environment that is not only lean and waste-free but also agile, adaptive, and technologically advanced. This integration underscores the evolutionary journey of manufacturing practices, bridging historical foundations with the transformative possibilities of the digital era.

Smart Business Models

Data-Driven Decision Making

Strategic insights: Smart business models make use of data to make strategic decisions.

Large-scale data collection and analysis helps organisations understand consumer behaviour, market trends, and operational effectiveness.

Real-Time Analytics: Organisations may make well-informed decisions quickly when real-time analytics are integrated. This flexibility is especially useful in corporate settings that are dynamic and changing quickly.

Customer-focused methods

Personalisation: Smart business models place a high value on analysing data to understand customers' needs. This insight enables individualised experiences through tailored services, product recommendations, or focused marketing campaigns.

Enhanced Customer Engagement: By utilising technology, companies may interact with clients across a variety of platforms, building closer bonds with them. Mobile apps, social media, and interactive platforms all help to create a smooth and customised client experience.

Improved supply chain visibility and predictive capabilities

End-to-end visibility across supply chains can be achieved by integrating IoT sensors, RFID technology and other tracking mechanisms to track goods throughout the manufacturing and delivery process. Data analytics applied to connected supply chain data enables predictive capabilities. Organisations can anticipate fluctuations in demand, optimise inventory levels and streamline logistics for more efficient and responsive operations.

Digital collaboration and strategic ecosystem partnerships

Collaboration platforms can be used to connect with suppliers, partners and other stakeholders in real time. Cloud tools and platforms facilitate seamless communication, document sharing and joint decision-making. The concept of business ecosystems is prominent, where organisations form strategic partnerships to create value beyond traditional supplier-customer relationships through diverse industry collaborations.

In today's data-driven business environment, organisations are redefining their approaches to embrace insights from analytics and support connected ecosystems. Smart business models use technology to improve decision-making and prioritise customer-centric innovation. Meanwhile, optimised supply chains and collaboration platforms support seamless interactions and strategic partnerships that enable continuous growth, innovation and agility in dynamic markets.

Applications in Supply Chain and Smart Contracts

Supply Chain Traceability: Blockchain provides end-to-end traceability of the supply chain. Every step of the chain, from purchasing raw materials to final delivery, can be recorded on the blockchain. This transparency increases accountability, reduces fraud, and ensures compliance with quality standards.

Smart contracts: Smart contracts are self-signed contracts whose terms are written directly in code, find implementation in IIoT electronically, and ensure the protection of contracts. For example, in supply chain management, smart contracts can process payments or trigger compliance with certain conditions.

5G Integration in IIoT

Improve connectivity: Integrating 5G technology into IIoT systems can improve connectivity. This high-speed, low-latency network enables seamless communication between devices, easy data exchange, and advanced functionality of IIoT applications.

Scalability: The scalability of 5G allows multiple devices to be connected simultaneously. This is especially useful in industrial environments where large numbers of sensors, actuators and devices need to communicate effectively.

Advances in Artificial Intelligence and Machine Learning

Predictive Analytics: Advances in Artificial Intelligence and Machine Learning can enable better predictive analytics in Industrial IoT applications. Improved algorithms can analyse historical data, predict equipment failure, improve production processes and improve overall performance.

Autonomous decision-making: AI-assisted decision-making has the potential to be more autonomous. Having IoT systems equipped with artificial intelligence can quickly make decisions based on complex data models and reduce the need for human intervention in daily processes.

Sustainability Measures

Environmental Monitoring: The Internet of Things promotes environmental monitoring in business. Sensors can track emissions, energy consumption and resource use, providing information for sustainable practices and compliance with environmental regulations.

Circular Economy Applications: Industrial IoT promotes circular economy practices by using resources efficiently, reducing waste and encouraging recycling. This is in line with the growing importance of a sustainable and environmentally friendly business.

The future of the Internet of Things business is determined by the integration of new technologies and new processes. Edge and cloud computing solve latency issues and increase efficiency, while blockchain provides security and transparency, especially in the supply chain. The convergence of 5G, advances in artificial intelligence and machine learning, and a focus on the potential to further fuel the growth of Industrial IoT are paving the way for a more efficient, smarter, and greener economy.

13.9 Review of IIoT Core Value Statements

The Internet of Things (IIoT) is a transformative force with many benefits that address the way business operates and grows. Analysis of key points shows the far-reaching impact on business processes:

IIoT integrates sensors and data analysis to provide rapid insight for good decision-making, control and efficient use of resources.

Cost Savings and Efficiency: IIoT enables energy management, product analysis and process simplification for good results thanks to decision-making data.

Quality Control and Process Improvement: Continuous Data Analysis Real-Time Monitoring in IIoT makes quality control, error detection and process improvement even easier.

Supply Chain Optimization: IIoT provides end-to-end visibility in the supply chain, improving product quality, transportation and improving collaboration with suppliers.

Improve Security and Compliance: IoT applications protect employees through real-time monitoring, detect potential threats and, as a rule, support compliance.

Customization and adaptability: The Internet of Things promotes productivity, allowing businesses to adjust production processes to meet specific needs and respond quickly to customer needs.

Data-Driven Decisions: IIoT produces large amounts of data, providing businesses with information for strategic planning, improving overall business performance and providing competitive advantage.

Customers and innovation: The development of smart products, improved customer connectivity through IoT connected devices, continuous innovation based on customer feedback and increased competition will help create a good business and consumer market.

13.10 The Role of IIoT in Shaping the Future of Industrial Processes

The role of IIoT extends beyond the present, shaping the trajectory of industrial processes for the future. Key aspects include:

Key areas include:

Technological Advances: The integration of the Internet of Things with new technologies such as artificial intelligence, blockchain and 5G is paving the way for unprecedented connectivity, security and efficiency.

Continuous Improvement: Enterprise IoT is a force for innovation and change, fostering a culture of continuous improvement and responsiveness to business changes.

Industry 4.0 transformation: IIoT is the foundation of the Industry 4.0 paradigm and promotes the integration of technology and physical processes to create a smart, connected environment with the business ecosystem.

Global Competitiveness: Businesses that embrace the Internet of Things will gain global competitiveness, attract the best talent, increase jobs and become leaders in their industries.

Sustainability: The Internet of Things contributes to sustainable development by saving energy, reducing waste and promoting resource management.

Resilience and Adaptability: IIoT provides businesses with the flexibility and adaptability they need to cope with uncertainty, disruption and rapid change in the global business environment.

In summary, the value proposition of Industrial IoT, from efficient operation to new equipment, demonstrates its important role in improving the system business standard. As businesses continue to adopt and integrate Industrial IoT solutions, connectivity, data-driven visibility and speed will become critical to growth and competitiveness in the future. The revolutionary power of Industrial IoT is driving the Industry 4.0 journey, enabling businesses to not only succeed today but also be prepared for the challenges and opportunities of tomorrow.

Industrial Internet of Things (IIoT) speaks to the meeting of conventional Manufacturing forms with cutting edge data innovation and network. Within the setting of IIoT, physical gadgets, machines, and frameworks are prepared with sensors, actuators, and connectivity, permitting them to gather, trade, and analyse information. This interconnected biological system empowers businesses to tackle real-time bits of knowledge, computerise forms, and improve by and large efficiency.

Key Components of IIoT:

Sensors and Actuators: Gadgets inserted with sensors capture real-world information, whereas actuators empower machines to reply to advanced commands.

Connectivity: IIoT depends on a strong network, regularly encouraged by remote systems, to empower consistent communication between gadgets and systems.

Data Analytics: Progressed analytics and machine learning calculations prepare the tremendous sums of information created by IIoT gadgets, giving significant experiences for decision-making.

Cloud Computing: Cloud framework is regularly utilised for storing and handling IIoT information, advertising adaptability and accessibility.

Significance of IIoT within the Oil, Chemical, and Pharmaceutical Industry:

The oil, chemical, and pharmaceutical businesses work in complex situations with rigid security and administrative prerequisites. The appropriation of IIoT in these segments brings forward a few transformative benefits:

Predictive Support and Resource Optimization:

IIoT empowers real-time checking of hardware wellbeing in oil refineries, chemical plants, and pharmaceutical fabricating units.

Predictive support calculations analyse information from sensors to expect hardware disappointments, minimising downtime and optimising resource performance.

Process Optimization and Control:

In chemical fabricating, IIoT permits for ceaseless observing of chemical forms, guaranteeing ideal conditions and quality control.

Pharmaceutical generation benefits from computerised control frameworks that keep up exact conditions for sedate manufacturing.

Supply Chain Management:

IIoT applications give perceivability into the supply chain for oil shipments, chemical conveyances, and pharmaceutical logistics.

Topic	Keynote
Prepare for Industry 4.0	Evaluation of readiness for Industry 4.0. Identification of challenges and uncertainties.
Strategy and Planning	Implementation of digital technology in existing processes. Exploration of new business models.
From Industry 3.0 to Industry 4.0	Evolution of technology business through different stages. Understanding the rise of the Industrial Internet of Things (IIoT).
Industrial Computerization	Historical context of automation in industries.
Limitations of Industry 3.0	Challenges faced in the previous industrial era.
Innovation	Importance of innovation in driving industry transformation.
Basic Principles of Industry 4.0	Fundamental principles guiding Industry 4.0 initiatives.
LEAN Production Systems	Adoption of lean methodologies in production processes.
Blockchain in Industrial IoT	Role of blockchain technology in enhancing security and transparency.
Review of IIoT Core Value Statements	Examination of core value propositions in Industrial IoT.

Topic	Keynote
The Role of IIoT in Shaping the Future	Impact of IIoT on the evolution of industrial processes.
Future Patterns and Advancements in IIoT Applications	Trends and innovations shaping the future of IIoT applications.

Summary:

This Chapter delves into the transformation from Industry 3.0 to Industry 4.0, highlighting the advancements in technology and the necessary preparations for this transition. It explores the historical context of Industry 3.0 and the emergence of Industry 4.0, emphasizing the role of industrial computerization and the limitations of the previous era. The chapter also discusses innovation, basic principles of Industry 4.0, lean production systems, blockchain's relevance in Industrial IoT, and the future patterns and advancements in IIoT applications.

Trivia Challenge: Industry 4.0

Welcome to the Industry 4.0 Trivia Challenge! Get ready to test your knowledge of the wacky world of Industry 4.0 with these hilarious multiple-choice questions. Each question is paired with some side-splitting facts and figures about Industry 4.0. Let us dive in and have a giggle!

1) What's the main objective of understanding the transformation from Industry 3.0 to Industry 4.0?

A) To reminisce about the good old days of Industry 3.0.

B) To prepare for Industry 4.0 by grasping its evolution.

C) To avoid Industry 4.0 altogether and stick with the familiar.

D) To explore the future of Industry 5.0 before it's even here.

2) How does Industry 4.0 shape global trends?

A) By resisting change and maintaining the status quo.

B) By implementing LEAN Production Systems to reduce waste.

C) By exploring different strategies and stages in its evolution.

D) By delving into IIoT Blockchain and its role in industrial IoT.

3) What's the LEAN Production System's favourite hobby?

A) Accumulating unnecessary inventory.

B) Minimizing waste and maximizing efficiency.

C) Taking long breaks and avoiding work.

D) Disrupting Industry 4.0 strategies with inefficiency.

4) Why did Industry 3.0 go out of style?

A) It couldn't keep up with Industry 4.0's cool gadgets.

B) Because it relied too heavily on manual labour and outdated processes.

C) Because it preferred strolls over fast-paced innovation.

D) Because it didn't want to evolve and embrace new technologies.

5) How does IIoT Blockchain fit into the future of industrial IoT?

A) By resisting change and sticking with traditional methods.

B) By exploring core value statements and future patterns.

C) By disrupting traditional industries with innovative technologies.

D) By leveraging blockchain technology to enhance data security and transparency.

6) What's Industry 4.0's favourite pastime?

A) Sticking to the same old routines and resisting change.

B) Embracing innovation and exploring new technologies.

C) Avoiding the future and clinging to the past.

D) Watching reruns of Industry 3.0's greatest hits.

7) How does Industry 4.0 prepare for the future?

A) By ignoring global trends and staying in its comfort zone.

B) By implementing LEAN Production Systems to reduce waste.

C) By embracing IIoT Blockchain and its transformative potential.

D) By avoiding IIoT applications and sticking with outdated technologies.

8) What's the LEAN Production System's favourite tool?

A) A chainsaw for cutting through inefficiencies.

B) A magnifying glass for examining waste.

C) A stopwatch for timing production processes.

D) A calculator for measuring cost savings.

9) Why did Industry 4.0 get invited to the innovation party?

    A) It knows how to have a good time and embrace change.

    B) Because it resisted change and stuck with outdated technologies.

    C) Because it understood the transformation journey from Industry 3.0.

    D) Because it wanted to reminisce about the good old days of Industry 3.0.

10) How does Industry 4.0 disrupt traditional industries?

    A) By embracing change and exploring new technologies.

    B) By clinging to outdated methods and resisting innovation.

    C) By implementing LEAN Production Systems to maximize efficiency.

    D) By avoiding IIoT Blockchain and its transformative potential.

Abbreviations:

Abbreviations	Terms
PI4	Prepare for Industry 4.0
SP	Strategy and Planning
I3I4	From Industry 3.0 to Industry 4.0
IC	Industrial Computerization
LI3	Limitations of Industry 3.0
IN	Innovation
BPI4	Basic Principles of Industry 4.0
LPS	LEAN Production Systems
BIIoT	Blockchain in Industrial IoT
RICVS	Review of IIoT Core Value Statements
TRISF	The Role of IIoT in Shaping the Future
FPIAIIoT	Future Patterns and Advancements in IIoT Applications

Chapter 14

Industrial IoT Applications & Use Cases

Learning Objective:

- Explore the importance of IIoT in transforming traditional manufacturing processes.

- Analyse the role of IIoT in enhancing operational productivity through real-time data analytics and process optimization.

- Examine the application of IIoT in different industries, such as healthcare, inventory management, quality control, and building automation.

- Identify the potential benefits and challenges associated with implementing IIoT solutions in industrial settings.

- Evaluate the impact of IIoT on various application domains, including oil, chemical, and pharmaceutical industries.

- Understand the implications of IIoT on the Fourth Industrial Revolution and its role in driving digital transformation across businesses.

- Explore emerging trends and advancements in IIoT technologies and their potential implications for future industrial applications.

14.0 IIoT Industrial Applications

The Industrial Internet of Things (IIoT) speaks to a progressive worldview within the integration of advanced innovations with Industrial forms. Unlike conventional Industrial setups, IIoT leverages interconnected gadgets, sensors, and advanced analytics to form smartly, data-driven environments inside different industrial divisions. This combination of physical and computerised domains permits for uncommon levels of industries, proficiency, and knowledge into manufacturing operations.

14.1 Key components of IIoT

Sensors and Gadgets: IIoT depends on an arrangement of sensors and gadgets inserted in apparatus, hardware, and mechanical situations. These sensors collect real-time information on different parameters, empowering a consistent stream of information.

Connectivity Arrangements: Vigorous network, both wired and remote, is fundamental for the consistent communication between IIoT gadgets. This network shapes the spine for the integration of dissimilar frameworks and encourages the trade of data.

Data Analytics: The huge volume of information produced by IIoT gadgets is handled utilising progressed analytics instruments. This information examination yields important bits of knowledge, empowering businesses to optimise forms, make educated choices, and foresee potential issues.

Automation and Control Frameworks: IIoT encourages the computerization of mechanical forms through the integration of savvy gadgets. This robotization upgrades accuracy, decreases human intercession, and leads to strides in general operational efficiency.

Cloud Computing: Cloud stages play an urgent part in IIoT by giving versatile and centralised capacity, preparing, and analytics capabilities. Cloud integration permits for the effective administration of tremendous sums of mechanical data.

14.2 Importance of IIoT in Changing Manufacturing Processes

Operational Productivity: IIoT drives operational productivity by giving real-time perceivability into mechanical forms. This perceivability permits businesses to distinguish bottlenecks, optimise workflows, and streamline operations for greatest productivity.

Predictive Upkeep: One of the key points of interest of IIoT is its capacity to empower prescient support. By ceaselessly observing the wellbeing of gear and apparatus, IIoT frameworks can foresee when support is required, decreasing downtime and amplifying the life expectancy of assets.

Quality Control: IIoT applications improve quality control by giving real-time bits of knowledge into the fabricating handle. Sensors and analytics can recognize surrenders, varieties, and deviations from quality benchmarks, guaranteeing that items meet rigid quality requirements.

Cost Lessening: Through robotization, prescient support, and preparation optimization, IIoT contributes to noteworthy decreases. Businesses can minimise vitality utilisation, diminish squander, and make data-driven choices that lead to more productive asset allocation.

Safety Advancements: IIoT plays a vital part in improving mechanical security. By checking frameworks and sensors, businesses can distinguish potential security dangers, track worker well-being, and react rapidly to crises, in this way making more secure working environments.

Supply Chain Optimization: IIoT empowers real-time following and observing of merchandise all through the supply chain. This perceivability upgrades supply chain productivity, decreases delays, and permits businesses to adapt rapidly to changes in request or disturbances within the supply chain.

Its integration into different segments is driven by the guarantee of expanded effectiveness, improved decision-making, and the capacity to form more dexterous and versatile mechanical biological systems. The significance of IIoT lies in its capacity to revolutionise conventional businesses, making them more associated, shrewdly, and responsive to the requests of the advanced era.

14.3 Healthcare

IIoT is utilised in healthcare to move forward quiet care, decrease costs, and increment effectiveness. For Illustration, IIoT can be utilised to -

Monitor patients' imperative signs remotely

Track the development of therapeutic gear and supplies

Manage stock of pharmaceuticals and other therapeutic supplies

Improve the productivity of clinic operations

Develop new diagnostic and treatment methods

14.3.1 Remote Persistent Monitoring:

Wearable Gadgets and Sensors

Definition: Inaccessible quiet observing in healthcare includes the utilisation of wearable gadgets prepared with sensors to gather and transmit wellbeing information from patients in real-time.

Examples: Wearables such as smartwatches, wellness trackers, and wellbeing checking gadgets that track crucial signs like heart rate, blood weight, and action levels.

Benefits:

Continuous Checking: Patients' well-being measurements are persistently checked, giving a comprehensive view of their well-being.

Early Location: Wearables can identify irregularities or changes in wellbeing parameters, empowering early intercession and preventive care.

Chronic Illness Administration: Especially profitable for overseeing constant conditions, guaranteeing proactive healthcare management.

Real-time Wellbeing Information Analytics:

Definition: Real-time wellbeing information analytics includes the examination of information collected from inaccessible quiet observing gadgets right away, permitting healthcare experts to determine significant insights.

Applications: Progressed analytics instruments prepare information to distinguish designs, patterns, and potential well-being risks.

Benefits:

Timely Intercession: Healthcare suppliers can get alarms for basic wellbeing occasions, empowering convenient intercession and decreasing crisis situations.

Personalised Care: Analysing real-time information permits for personalised treatment plans, altering care methodologies based on person wellbeing patterns.

Efficient Asset Allotment: Healing centres can optimise asset allotment by prioritising patients who require prompt attention.

14.3.2 Asset Following in Hospitals:

1. Gear Management:

Definition: Resource following in healthcare includes the utilisation of IoT gadgets and following frameworks to screen the area and status of restorative gear inside a hospital.

Examples: RFID labels, sensors, and following computer program coordinates with therapeutic gadgets and equipment.

Benefits:

Reduced Look Times: Healing centres can rapidly find and recover fundamental gear, decreasing the time spent looking for resources.

Preventive Support: Resource following encourages proactive upkeep plans, minimising gear downtime and guaranteeing solid operation.

Optimised Workflow: Effective administration of restorative hardware progresses workflow and improves by and large clinic efficiency.

2. Stock Control:

Definition: IIoT empowers real-time following of restorative supplies and pharmaceuticals, guaranteeing exact and proficient stock management.

Applications: RFID labels, sensors, and stock administration computer program coordinates with capacity systems.

Benefits:

Avoiding Stock Outs: Real-time stock control makes a difference to avoid deficiencies of basic supplies, guaranteeing that fundamental things are continuously available.

Reduced Wastage: Exact following permits for superior control of close dates, decreasing the chance of terminated medicines or supplies.

Cost Investment funds: Productive stock administration leads to taking a toll reserve funds through diminished squander and ideal stock levels.

IIoT applications in healthcare, such as inaccessible understanding checking and resource following, are changing the industry by giving real-time bits of knowledge, making strides in understanding care, and improving operational proficiency inside healing centres. These advances clear the way for a more associated and responsive healthcare biological system.

14.3.3 Predictive Maintenance:

Observing Gear Health:

Definition: Prescient upkeep in control plants includes the utilisation of Mechanical Web of Things (IIoT) advances to ceaselessly screen the wellbeing and execution of basic equipment.

Implementation: Sensors and checking gadgets are introduced on key components, such as turbines, generators, and pumps, to gather real-time information on operational parameters.

Benefits:

Condition Checking: Nonstop checking permits for real-time evaluation of hardware condition, distinguishing potential issues some time recently they escalate.

Increased Uptime: Early location of irregularities empowers proactive support, lessening the chance of startling breakdowns and maximising plant uptime.

Extended Hardware Life expectancy: Prescient support phones contribute to amplifying the life expectancy of costly and basic assets.

Early Discovery of Anomalies:

Definition: IIoT encourages the early location of peculiarities in control plant gear by analysing information designs and deviations from ordinary working conditions.

Analytics Apparatuses: Progressed analytics apparatuses prepare information from different sensors to recognize abnormal trends or behaviours that will demonstrate approaching issues.

Benefits:

Risk Moderation: Early peculiarity location permits for convenient intercession, diminishing the hazard of gear disappointments, mischances, and impromptu downtime.

Cost Reserve funds: Proactive support based on irregularity location is more cost-effective than responsive repairs, minimising by and large upkeep costs.

Enhanced Security: Recognizing potential security risks in development guarantees a more secure working environment for plant administrators and support personnel.

Energy Efficiency:

Keen Grids:

Definition: Savvy frameworks use IIoT innovations to form cleverly interconnected control dissemination systems that empower two-way communication between utilities and end-users.

Components: Savvy metres, sensors, and communication infrastructure enable real-time checking and control of power distribution.

Benefits:

Demand Reaction: Keen frameworks permit utilities to reply powerfully to vacillations in power request, optimising vitality distribution.

Reduced Vitality Misfortunes: Progressed checking and control components minimise vitality misfortunes amid conveyance, progressing in general efficiency.

Integration of Renewable Vitality: Savvy networks encourage the integration of renewable vitality sources by productively overseeing variable control generation.

Optimization of Vitality Consumption:

Definition: IIoT empowers the optimization of vitality utilisation inside control plants by analysing information on vitality utilisation and executing proficiency measures.

Applications: Progressed sensors and analytics instruments track vitality utilisation designs and recognize zones for improvement.

Benefits:

Resource Optimization: IIoT makes a difference when control plants distinguish wasteful aspects in vitality utilisation, permitting for the optimization of asset usage.

Cost Diminishment: By optimising vitality utilisation, control plants can decrease operational costs and upgrade generally profitability.

Environmental Affect: Made strides vitality effectiveness contributes to a decrease in carbon emanations and underpins supportability goals.

The integration of IIoT in control plants is revolutionising the industry by empowering prescient upkeep hones, early peculiarity discovery, and optimising energy productivity. These applications improve the unwavering quality, security, and supportability of the control era, stamping a critical headway within the advancement of the vitality sector.

14.4 Inventory Administration & Quality Control

IIoT is utilised in stock administration and quality control to progress precision, proficiency, and item quality. For illustration, IIoT can be utilised to: -

Track the development of stock in real-time

Monitor stock levels and distinguish potential stock outs

Automate quality control processes

Identify and review inadequate items rapidly

14.4.1 Real-time Inventory Tracking:

RFID Technology:

Definition: Real-time stock following with RFID (Radio-Frequency Recognizable proof) innovation includes the utilisation of RFID labels and peruses to screen the development and status of stock things in genuine time.

Implementation: RFID labels are connected to each thing, and RFID perusers put deliberately all through the office capture information, upgrading the stock database instantaneously.

Benefits:

Accuracy: RFID innovation gives precise and real-time data on stock levels, minimising mistakes related with manual following methods.

Efficiency: Computerised information capture and real-time overhauls streamline stock administration forms, diminishing the time and exertion required for manual stock checks.

Visibility: Total perceivability into stock developments empowers way better decision-making, arrange fulfilment, and request forecasting.

14.4.1.1 Supply Chain Optimization:

Definition: Real-time stock following contributes to supply chain optimization by guaranteeing perceivability and control over stock levels all through the whole supply chain.

Applications: Integration of real-time stock information with supply chain administration frameworks permits for energetic alterations and improvements.

Benefits:

Reduced Stockouts: Real-time experiences into stock levels empower proactive renewal, minimising the chance of stock outs and guaranteeing a persistent supply chain.

Lower Holding Costs: Optimising stock levels based on real-time requests decreases holding costs related to abundance stock.

Improved Client Fulfilment: Effective supply chain administration guarantees convenient arrangement fulfilment, upgrading client satisfaction.

14.4.1.2 Quality Control in Manufacturing:

Ceaseless Observing with Sensors:

Definition: Quality control in fabricating includes the ceaseless observing of generation forms utilising sensors to guarantee that items meet predefined quality standards.

Sensor Sorts: Different sensors, such as vision sensors, weight sensors, and temperature sensors, are coordinated into fabricating hardware to screen and degree basic parameters.

Benefits:

Early Location of Deviations: Sensors distinguish deviations from quality parameters in genuine time, permitting for quick remedial action.

Consistent Item Quality: Nonstop checking guarantees that items reliably meet quality benchmarks, diminishing inconstancy within the fabricating process.

Process Optimization: Bits of knowledge from sensor information can be utilised to optimise fabricating forms for superior effectiveness and quality.

14.4.1.3 Mechanised Deformity Detection:

Definition: Robotized deformity location includes the utilisation of AI-powered frameworks and machine vision to recognize abandonment and abnormalities in fabricated products.

Implementation: Cameras and imaging frameworks capture pictures of items, and machine learning calculations analyse the pictures to recognize defects.

Benefits:

Increased Precision: Robotized frameworks can identify abandons with a tall level of exactness, minimising the chances of neglecting unobtrusive imperfections.

Efficient

Review: Computerised imperfection discovery is speedier than manual review, contributing to expanded generation efficiency.

Reduced Scrap and Adjust: Early discovery of surrenders permits to provoke remedial activity, diminishing the requirement for broad revamp and minimising waste.

The execution of real-time stock following RFID innovation and progressed quality control measures in fabricating essentially improves productivity, precision, and in general operational brilliance in stock administration and production processes. These applications are significant for businesses endeavouring to preserve tall item quality, streamline supply chains, and optimise their fabricating operations.

14.4.2 Plant Safety and Security:

Paramount contemplations in mechanical settings.

Encompassing a extent of measures and innovations outlined to secure faculty.

Assets.

The environment.

The integration of progressed innovations plays a vital part in improving security conventions and bracing security measures inside mechanical plants.

Safety Measures:

Risk Assessments:

Definition: Hazard evaluations include distinguishing and assessing potential risks inside the plant environment to execute preventive measures.

Implementation: Utilising strategies such as HAZOP (Danger and Operability Ponder) to efficiently evaluate and relieve risks.

Benefits:

Proactive Security Arranging: Distinguishing potential dangers permits for the usage of proactive security measures.

Reduced Mischance Rates: By tending to dangers some time recently they heighten, the probability of mishaps is essentially reduced.

Compliance: Adherence to security controls and benchmarks is guaranteed through comprehensive chance assessments.

Safety Preparing and Awareness:

Definition: Security preparing programs teach the plant work force on security conventions, crisis methods, and the right utilisation of security equipment.

Implementation: Conducting customary security drills, workshops, and giving get to instructive resources.

Benefits:

Preparedness: Well-trained personnel are way better arranged to reply viably to emergencies.

Reduced Human Blunder: Preparing upgrades mindfulness, decreasing the probability of human mistakes that seem to lead to accidents.

Cultural Move: Cultivates a safety-conscious culture where representatives prioritise and effectively take an interest in keeping up a secure working environment.

14.4.3 Personal Defensive Equipment (PPE):

Definition: Individual defensive hardware incorporates adaptations such as protective caps, gloves, security glasses, and other specialised clothing planned to secure specialists from particular hazards.

Implementation: Giving and ordering the utilisation of fitting PPE based on work parts and potential risks.

Benefits:

Injury Avoidance: PPE acts as a physical obstruction, anticipating wounds from risks like falling objects, chemicals, or electrical shocks.

Compliance with Directions: Guarantees compliance with word related security guidelines and regulations.

Employee Well-being: Illustrates a commitment to representative well-being and cultivates a more secure work environment.

Security Measures:

Perimeter Security:

Definition: Edge security includes measures to secure the external boundaries of the plant, anticipating unauthorised access.

Implementation: Fencing, get to doors, reconnaissance cameras, and movement sensors to screen and control access.

Benefits:

Deterrence: Obvious border security measures prevent unauthorised people from endeavouring to breach the facility.

Access Control: Controls the stream of individuals and vehicles, guaranteeing that as it were authorised staff enter confined areas.

Incident Reaction: Empowers quick reactions to security breaches through real-time checking and alerts.

Security Reconnaissance Systems:

Definition: Security reconnaissance frameworks utilise cameras and sensors to screen plant premises, giving visual documentation and analysis.

Implementation: Conveying CCTV cameras, infrared sensors, and video analytics for real-time monitoring.

Benefits:

Incident Examination: Observation film helps in examining episodes, mischances, or security breaches.

Deterrence and Anticipation: The nearness of observation frameworks disheartens criminal exercises and advances a more secure environment.

Remote Checking: Permits for inaccessible observing of security bolsters, upgrading in general situational awareness.

Access Control Systems:

Definition: Get to control frameworks direct and oversee section and exit focuses, confining get to authorised personnel.

Implementation: Key card frameworks, biometric confirmation, and electronic get to controls.

Benefits:

Unauthorised Avoidance: Guarantees that as it were people with appropriate accreditations pick up to particular areas.

Audit Trails: Gives a record of passage and exit, supporting in responsibility and occurrence investigations.

14.4.4 Facility Management

IIoT is utilised in office administration to make strides effectiveness, diminish costs, and make strides inhabitant consolation and security. For case, IIoT can be utilised to -

Monitor and control HVAC frameworks in genuine time

Automate upkeep schedules

Reduce vitality consumption

Improve discuss quality

Provide inhabitants with real-time data on the status of the facility

14.5 Building Automation

HVAC Frameworks Optimization:

Definition: Building mechanisation for HVAC (Warming, Ventilation, and Discuss Conditioning) includes the utilisation of keen advances to optimise the execution and vitality proficiency of HVAC systems.

Implementation: Sensors and robotization controllers screen natural conditions and alter HVAC settings in genuine time based on inhabitants, climate conditions, and vitality consumption.

Benefits:

Energy Proficiency: HVAC optimization diminishes vitality utilisation by guaranteeing that warming and cooling frameworks work at ideal levels.

Cost Investment funds: Keen HVAC frameworks offer assistance in decreasing utility costs by maintaining a strategic distance from superfluous vitality utilisation amid periods of moo occupancy.

Occupant Consolation: Computerization guarantees that indoor temperatures are kept up at comfortable levels, improving the general inhabitant experience.

Lighting and Vitality Management:

Definition: Building computerization expands to lighting and vitality administration by coordination keen lighting frameworks and controls to optimise vitality usage.

Implementation: Movement sensors, sunshine gathering frameworks, and keen lighting controls are coordinates to naturally alter lighting levels based on inhabitants and common light availability.

Benefits:

Energy Preservation: Mechanised lighting control decreases vitality utilisation by turning off lights in vacant zones and altering brightness levels as needed.

Enhanced Supportability: Effective vitality administration contributes to a facility's supportability objectives, lessening its natural impact.

Operational Proficiency: Robotization disentangles lighting administration, diminishing the workload on office administration faculty and guaranteeing reliable energy-efficient practices.

14.6 Security and Surveillance

Video Analytics:

Definition: Video analytics includes the utilisation of AI-powered innovations to analyse video information from observation cameras, giving bits of knowledge into security and operational aspects.

Applications: Video analytics can distinguish unordinary exercises, track objects or people, and give real-time cautions for security incidents.

Benefits:

Intrusion Location: Video analytics can distinguish unauthorised get to or suspicious behaviour, activating quick security responses.

Operational Bits of knowledge: Past security, video analytics can give experiences into foot activity, inhabitants designs, and other operational aspects.

Efficient Checking: Robotization diminishes the requirement for steady manual checking, permitting the security work force to centre on basic situations.

Get to Control Systems:

Definition: Get to control frameworks utilise progressed innovations to control and screen passage to buildings or particular zones inside a facility.

Components: Get to control frameworks may incorporate key cards, biometric verification, and keen locks, all overseen through centralised control systems.

Benefits:

Enhanced Security: Get to control limits unauthorised get to, decreasing the hazard of robbery, vandalism, or other security breaches.

Audit Trails: These frameworks give nitty gritty review trails, permitting office supervisors to track and survey events.

Flexibility: Get to control frameworks offer adaptable get to administration, permitting for the customization of consents based on parts and responsibilities.

Facility administration, upheld by building robotization and progressing security measures, not as it were, makes strides operational proficiency but moreover improves the security, security, and maintainability of the built environment. These advances contribute to making shrewdly, responsive, and secure offices that meet the advancing needs of tenants and owners.

IIoT is utilised in different mechanical applications, counting healthcare, control plants, stock administration and quality control, plant security and security, and office administration. IIoT can progress effectiveness, efficiency, quality, and security in these applications.

AR and VR are too utilised with IIoT to create unused security applications. These applications can offer assistance to prepare specialists, give them real-time data on dangers, and offer assistance to dodge risks.

IIoT could be a transformative innovation that altogether impacts the mechanical segment. By actualizing IIoT arrangements, organisations can move forward their foot line and make a more secure and more productive workplace.

14.7 Application Domains: Oil, Chemical and Pharmaceutical Industry

14.7.1 Oil Industry: Leveraging IIoT for Efficiency and Safety

Predictive Maintenance for Equipment:

Monitoring and Keeping up Oil Rigs:

IIoT Sensors: Oil rigs are prepared with sensors that screen the wellbeing of basic components such as pumps, engines, and boring equipment.

Real-time Information Examination: IIoT analytics handle information in genuine time, recognizing irregularities and foreseeing potential hardware disappointments some time recently.

Reduced Downtime: Prescient upkeep permits for convenient intercessions, lessening spontaneous downtime and expanding the life expectancy of costly penetrating equipment.

Pipeline Astuteness Monitoring:

Sensor Systems: IIoT-enabled sensors are conveyed along pipelines to screen components like weight, temperature, and corrosion.

Early Location: Persistent checking empowers the early location of pipeline issues such as spills or basic weaknesses.

Preventive Measures: Prescient analytics recommend support intercessions to anticipate pipeline disappointments, guaranteeing the keenness of the whole system.

Resource Following and Optimization:

Tracking Oil Tankers and Cargo:

GPS and RFID Innovations: IIoT coordinating GPS and RFID innovations to track the area and condition of oil tankers in genuine time.

Cargo Checking: Sensors give bits of knowledge into variables like temperature, weight, and vibration, guaranteeing the keenness of transported oil.

Route Optimization: IIoT analytics optimise transportation courses, minimising fuel utilisation and guaranteeing opportune deliveries.

Optimization of Refinery Operations:

Sensor-driven Bits of knowledge: IIoT sensors in refineries collect information on gear execution, vitality utilisation, and natural factors.

Process Optimization: Real-time analytics recognize openings for preparation optimization, making strides effectiveness and decreasing vitality consumption.

Resource Allotment: IIoT-driven bits of knowledge direct ideal assignment of assets, generally improving refinery performance.

Security and Natural Monitoring:

Real-time Observing of Gas Emissions:

Gas Sensors: IIoT sends gas sensors all through the oil generation and refining to screen emissions.

Immediate Alarms: Irregularities trigger quick alarms, permitting quick reaction to moderate natural impact.

Regulatory Compliance: Persistent observing guarantees compliance with natural directions and minimises the chance of fines.

Emergency Reaction and Emergency Management:

Connected Crisis Frameworks: IIoT coordinating crisis reaction frameworks to identify and react to basic situations.

Predictive Analytics: Prescient capabilities help in foreseeing potential emergency scenarios, empowering proactive measures.

Communication Systems: IIoT improves communication amid crises, encouraging coordination between on-site faculty and off-site response teams.

The application of IIoT within the oil industry revolutionises conventional phones, bringing almost prescient support, resource optimization, and improved security measures. The real-time experiences given by IIoT innovations not as it were to move forward operational productivity but to contribute to the industry's commitment to natural maintainability and security. The oil industry's selection of IIoT speaks to a critical walk towards more astute, more versatile, and feasible vitality generation.

14.7.2 Chemical Industry: Transforming Operations with IIoT

Handle Optimization and Control:

Real-time Checking of Chemical Processes:

Sensors and IIoT Integration: Chemical plants send sensors over different stages of the generation handle, giving real-time information on temperature, weight, and chemical reactions.

Predictive Analytics: IIoT analytics prepare the collected information, advertising bits of knowledge to handle effectiveness, recognizing bottlenecks, and guaranteeing ideal conditions.

Automated Alterations: Robotized control frameworks, driven by IIoT bits of knowledge, make real-time alterations to chemical forms, optimising efficiency and minimising waste.

Automated Control Frameworks for Quality Assurance:

Quality Sensors: IIoT-integrated quality sensors screen key parameters to guarantee the generation of chemicals meets predefined quality standards.

Closed-loop Control: Robotized control frameworks utilise criticism from sensors to create momentary alterations, keeping up steady item quality.

Reduced Changeability: IIoT-driven control frameworks minimise varieties, upgrading the unwavering quality and consistency of chemical production.

Supply Chain Management:

Tracking and Observing Chemical Shipments:

GPS and RFID Integration: IIoT joins GPS and RFID innovations to track chemical shipments in transit.

Condition Checking: Sensors on holders give real-time information on components like temperature, stickiness, and stun, guaranteeing the keenness of chemical products.

Route Optimization: IIoT analytics optimise transportation courses, diminishing travel time and upgrading the general productivity of the supply chain.

Inventory Administration and Request Forecasting:

Smart Stock Frameworks: IIoT-enabled stock frameworks screen stock levels, naturally activating reorder focuses based on real-time utilisation data.

Demand Forecast: Prescient analytics analyse chronicled information and advertise patterns, encouraging exact request forecasting.

Reduced Stock outs and Abundance Stock: IIoT-driven bits of knowledge anticipate stock outs and overstock circumstances, optimising working capital and making strides supply chain resilience.

Security Compliance and Danger Detection:

Continuous Observing for Perilous Conditions:

Gas and Chemical Sensors: IIoT coordinating sensors to screen discuss quality, distinguish spills, and identify possibly dangerous conditions within the chemical plant.

Immediate Alarms: Peculiarities trigger quick alarms, empowering fast reaction to moderate dangers and avoid accidents.

Worker Security: Ceaseless checking improves specialist security by giving early notices for clearing or defensive measures.

Compliance with Security Regulations:

Regulatory Compliance Frameworks: IIoT guarantees adherence to security directions by persistently checking and recording processes.

Audit Path: Nitty gritty records of security parameters, upkeep exercises, and crisis reactions bolster compliance audits.

Enhanced Detailing: IIoT-generated reports encourage straightforward communication with administrative specialists, illustrating the industry's commitment to safety.

The integration of real-time observing, Industrial controls, and prescient analytics positions the chemical industry to meet the advancing requests of quality confirmation, operational productivity, and security measures. IIoT proceeds to drive advancement in chemical fabricating, contributing to a more maintainable, versatile, and carefully progressed industry.

14.8.3 Pharmaceutical Industry: Elevating Operations through IIoT

Fabricating Prepare Optimization:

Real-time Observing of Pharmaceutical Production:

Sensors in Generation Lines: IIoT coordinating sensors into pharmaceutical fabricating gear to screen basic parameters such as temperature, weight, and blending ratios.

Continuous Information Stream: Real-time information gushing permits for persistent observing of generation forms, guaranteeing consistency and adherence to predefined standards.

Quality Experiences: IIoT analytics give experiences into fabricating effectiveness, distinguishing zones for optimization and keeping up the most noteworthy quality standards.

Quality Control and Assurance:

Integrated Quality Sensors: IIoT consolidates sensors to screen the quality qualities of pharmaceutical items, counting composition, virtue, and consistency.

Automated Quality Checks: IIoT-driven mechanised frameworks conduct real-time quality checks, guaranteeing that items meet administrative necessities and inner quality benchmarks.

Reduced Inconstancy: Mechanization through IIoT minimises varieties within the fabricating handle, upgrading the unwavering quality and reproducibility of pharmaceutical products.

Supply Chain Visibility:

Tracking and Following Pharmaceutical Products:

Serialized Track-and-Trace Frameworks: IIoT empowers the usage of serialised tracking systems to follow the complete travel of pharmaceutical items from fabricating to distribution.

Real-time Area Observing: GPS-enabled trackers give real-time area upgrades, guaranteeing exact following and lessening the hazard of counterfeiting.

Improved Review Administration: In case of reviews, IIoT encourages exact distinguishing proof and expulsion of influenced items from the supply chain, minimising potential risks.

Cold Chain Observing for Temperature-sensitive Products:

Temperature Sensors: IIoT coordinating temperature sensors into bundling and capacity units, guaranteeing the keenness of temperature-sensitive pharmaceuticals.

Real-time Observing: Ceaseless observing of temperature conditions amid transportation and capacity anticipates deviations that seem to compromise item efficacy.

Automated Cautions: IIoT-driven frameworks trigger cautions for deviations from indicated temperature ranges, permitting prompt remedial activities to protect item quality.

Administrative Compliance and Information Security:

Ensuring Compliance with Administrative Standards:

Documentation and Announcing: IIoT encourages mechanised documentation of fabricating and dispersion forms, guaranteeing compliance with administrative standards.

Audit Trails: Point by point review trails created by IIoT frameworks give a straightforward record of exercises, supporting administrative reviews and inspections.

Real-time Compliance Checking: Nonstop observing guarantees that fabricating hones adjust with advancing administrative requirements.

Data Security Measures for Touchy Information:

Secure Information Transmission: IIoT frameworks actualize encryption and secure communication conventions to defend the transmission of touchy data.

Access Controls: IIoT stages implement get to controls, limiting delicate data as it were to authorised personnel.

Cybersecurity Measures: Continuous cybersecurity measures, counting standard upgrades and risk checking, ensure against data breaches and unauthorised access.

The application of IIoT within the pharmaceutical industry propels fabricating forms, guarantees item quality, and generally improves supply chain perceivability. By grasping real-time checking, mechanisation, and compliance measures, the pharmaceutical division leverages IIoT to meet administrative benchmarks, keep up item astuteness, and reinforce its position in giving secure and compelling healthcare arrangements. The proceeded integration of IIoT innovations positions the pharmaceutical industry at the cutting edge of computerised change, cultivating advancement and supportability.

Challenges and Solutions in Industrial IoT (IIoT) Implementation

Tending to Security Concerns:

Cybersecurity Measures in Mechanical IoT:

Challenge: The interconnected nature of IIoT gadgets and frameworks makes vulnerabilities, uncovering mechanical forms to potential cyber threats.

Solution:

Firewall and Interruption Location Frameworks: Executing vigorous firewalls and interruption location frameworks improves the security border, avoiding unauthorised access.

Regular Security Reviews: Conducting occasional security reviews makes a difference to distinguish and address vulnerabilities within the IIoT infrastructure.

Endpoint Security: Securing person gadgets with upgraded antivirus program and customary fix administration shields against cyber threats.

Data Encryption and Privacy:

Challenge: IIoT includes the transmission of delicate information, and without appropriate encryption, this information is vulnerable to capture attempts and unauthorised access.

Solution:

End-to-End Encryption: Utilising end-to-end encryption guarantees that information remains secret amid transmission, securing it from interception.

Secure Key Administration: Executing secure key administration homes upgrades the adequacy of encryption and guarantees the judgement of communication channels.

Privacy Approaches: Building up and upholding security arrangements clarifies how information is collected, put away, and utilised, cultivating straightforwardness and compliance with information security regulations.

Overcoming Integration Challenges:

Interoperability of IIoT Gadgets and Systems:

Challenge: The differing run of IIoT gadgets and frameworks regularly ruins consistent communication and interoperability, driving to integration challenges.4

Solution:

Adoption of Standard Communication Conventions: Grasping broadly acknowledged communication conventions such as MQTT or CoAP guarantees compatibility between gadgets and systems.

Middleware Arrangements: Actualizing middleware stages that act as mediators between distinctive conventions and frameworks encourages smooth communication.

Open Measures: Empowering the appropriation of open measures advances compatibility and interoperability over different IIoT components.

Standardisation for Consistent Integration:

Challenge: Need of standardisation in IIoT conventions and information designs can hinder the integration of gadgets and frameworks from diverse manufacturers.

Solution:

Industry Standards Development: Collaborative endeavours to set up and follow industry-specific guidelines improve interoperability and ease integration.

IoT Stages: Utilising IoT stages that back numerous conventions and give deliberation layers rearranges integration.

Consortium Interest: Dynamic interest in industry consortia centred on IIoT measures permits organisations to contribute to and advantage from standardisation efforts.

Robust cybersecurity measures, encryption homes, and adherence to measures guarantee the assurance of mechanical frameworks and information. Moreover, endeavours toward interoperability and standardisation contribute to the consistent integration of assorted IIoT gadgets, cultivating an associated and collaborative mechanical environment. As the IIoT scene proceeds to advance, a proactive approach to security and interoperability will be vital for opening the complete potential of this transformative innovation.

In summary, the success stories and lessons learned from these case studies illustrate that thoughtful integration of IIoT technologies can yield tangible benefits, fostering innovation, sustainability, and competitiveness in the evolving industrial landscape.

Use Cases

Optimising Crop Management Using the Internet of Things (IoT)

Overview: A leading agricultural company implemented an Internet of Things (IoT) solution to optimise crop management practices on farms.

Business Model: The solution used a pay-as-you-go and subscription-based commercial model.

Implementation: Sensors monitored key farm metrics such as soil moisture, weather conditions and crop health in real time. Farmers subscribed to the service and paid based on the acreage managed.

Results: Farmers achieved better crop yields with precision irrigation and inputs based on real-time information. Resource efficiency has also increased through data-driven decision making.

Lessons Learned: Customizing IoT solutions for specific industry needs has proven critical to adoption and success. A flexible pricing model further supported user adoption.

Case study: Predicting maintenance of equipment in manufacturing

Overview: A leading automotive manufacturer deployed an IoT platform for predictive maintenance of equipment across manufacturing plants.

Business model: The solution followed a results-based business structure.

Implementation: Sensors continuously monitor equipment performance metrics to predict maintenance requirements before failures occur. Payments associated with reduced unplanned downtime and extended asset life.

Results: The plant achieved a 30% decrease in unplanned shutdowns along with a 20% increase in overall line efficiency.

Best Practices: Clearly defining quantifiable outcomes up front has proven critical to outcome-based models. Regular performance reviews ensured ongoing value delivery.

Case Study: Optimising Logistics Operations Using the Internet of Things

Overview: A leading logistics provider implemented an IoT platform for real-time shipment visibility and process optimization.

Business Model: The solution has adopted a subscription-based commercial model.

Implementation: GPS-enabled trackers monitored the location and status of shipments being transported. Subscribers gained access to real-time tracking data and analytics.

Results: Route optimization reduced delivery times by 15% while improving inventory visibility and supply chain efficiency.

Lessons Learned: IoT is transforming operational processes, improving visibility and efficiency across the value chain. Subscription models are consistent with continuous benefits realisation.

Case Study: Automation of Building Management with IoT

Overview: A major real estate company deployed an IoT platform for centralised management of building automation functions.

Business model: The solution followed the Platform as a Service (PaaS) business structure.

Implementation: The IoT platform provided unified oversight of HVAC, lighting, security and more. Customers have subscribed to automation services on PaaS.

Results: Buildings optimised HVAC operations through IoT and reduced energy consumption by 25%. Tenant satisfaction and operational efficiency have also improved.

Best Practices: Comprehensive training and support ensured user acceptance. Iterative improvements based on feedback improved long-term success.

These case studies demonstrate the diverse applications and successes of IoT across industries. Tailored implementations using flexible business models drive efficiency, cost savings and transformational impact through real-time process optimization and data-driven insights. Lessons emphasise the importance of adaptability, measurable results and sustained engagement.

The Industrial Internet of Things (IIoT) presents a rapidly growing market opportunity with the potential to transform industries. Businesses that can successfully develop and implement IIoT solutions will be well positioned to compete in the future.

Several IIoT business models show promise. Partnerships with technology companies and system integrators can reduce upfront investment and complexity for IIoT projects. Some businesses are developing IIoT ecosystems encompassing hardware and software vendors, data analytics providers, and application developers to create comprehensive, integrated solutions for customers. IIoT also enables new markets, such as IIoT-enabled products and services from industrial equipment companies and data analytics services directly targeting the IIoT sector.

As the IIoT market matures, more innovative and disruptive business models are likely to emerge. Successful implementation of IIoT stands to yield tangible benefits including improved efficiency, cost savings, and enhanced operations. However, security concerns, interoperability challenges, and the need for scalable, compliant systems require careful consideration and innovative solutions using best practices for encryption, standardisation, and modular architecture to ensure long-term success.

The integration of big data analytics and software-defined networking further enhances IIoT ecosystems' capabilities. Machine learning and data science provide predictive insights, quality control, and supply chain optimizations. Programming languages increase versatility in IIoT analytics. Hadoop offers a scalable data management solution, while security, scalability, edge analytics, artificial intelligence integration, and blockchain for data security underscore IIoT's dynamic nature.

Looking ahead, edge computing, artificial intelligence, and blockchain are anticipated to assume increasingly important roles by addressing challenges and unlocking new possibilities. Data centre networks, edge data centres, hybrid cloud architectures, and fusion of software-defined networking with IIoT analytics showcase ongoing evolution.

In conclusion, the Industrial Internet of Things represents the ongoing digital revolution's transformation of industries and redefinition of operational paradigms. IIoT technologies enhance efficiency, productivity, and open new fronts for innovation and competitiveness. As industries continue adapting and harnessing IIoT's power, the journey toward a connected, intelligent, efficient industrial future promises unprecedented advancements and growth through technology's catalytic role.

Key Notes:

Topic	Keynotes
Key components of IIoT	IIoT relies on sensors and devices embedded in machinery, equipment, and industrial environments to collect real-time data on various parameters. This data enables a continuous flow of information.
Importance of IIoT in Changing Manufacturing Processes	IIoT enhances operational productivity by providing real-time visibility into industrial processes, enabling businesses to identify bottlenecks, optimize workflows, and streamline operations for maximum efficiency.
Healthcare	IIoT is utilized in healthcare to improve patient care, reduce costs, and increase efficiency through remote monitoring, equipment tracking, inventory management, operational efficiency improvements, and new diagnostic and treatment methods.
Inventory Administration & Quality Control	IIoT facilitates real-time inventory tracking, supply chain optimization, quality control in manufacturing, automated defect detection, plant safety and security, personal protective equipment integration with HR systems, augmented reality (AR) applications, virtual reality (VR) for safety, and facility management.
Building Automation	IIoT plays a significant role in building automation by enabling smart building systems that optimize energy usage, enhance security, improve comfort, and streamline facility management.
Recap of Key IIoT Application Domains	Various application domains have emerged in the rapidly evolving landscape of IIoT, transforming the way industries operate. Key domains include oil, chemical, and pharmaceutical industries.
Impact of IIoT on the Fourth Industrial Revolution	IIoT is a key driver of the Fourth Industrial Revolution (4IR), integrating digital technologies into all aspects of business and revolutionizing manufacturing, healthcare, and other industries.

Trivia Challenge: Industry 4.0

Welcome to the Industry 4.0 Trivia Challenge! Get ready to test your knowledge of the wacky world of Industry 4.0 with these hilarious multiple-choice questions. Each question is paired with some side-splitting facts and figures about Industry 4.0. Let us dive in and have a giggle!

1) What's the Industrial IoT's favourite multitasking skill?

A) Juggling coffee mugs while typing emails.

B) Changing manufacturing processes and revolutionizing healthcare.

C) Balancing on a unicycle while solving equations.

D) Filing paperwork while conducting security surveillance.

2) How does Industrial IoT revolutionize inventory administration?

A) By creating chaos and confusion in warehouses.

B) By optimizing processes and enhancing efficiency.

C) By hiding inventory and playing hide-and-seek with workers.

D) By increasing downtime and disrupting workflow.

3) What's the preferred pastime of Industrial IoT in the pharmaceutical industry?

A) Making prescription drugs taste like candy.

B) Enhancing efficiency and optimizing processes.

C) Playing doctor and diagnosing illnesses.

D) Creating chaos in the laboratory and spilling chemicals.

4) What's the Industrial IoT's secret weapon in quality control?

A) An army of robots with laser vision.

B) A magnifying glass for scrutinizing every detail.

C) A magic wand for instantly fixing defects.

D) A blindfold for ignoring quality issues.

5) How does Industrial IoT contribute to building automation?

A) By randomly changing thermostat settings.

B) By optimizing processes and reducing energy consumption.

C) By creating chaos and confusion in smart buildings.

D) By painting walls and hanging curtains.

6) What's Industrial IoT's favourite bedtime story?

A) The Adventures of Data Security and Surveillance.

B) The Mystery of Inventory Administration.

C) The Tale of Process Optimization and Efficiency.

D) The Legend of Quality Control and Defect Detection.

7) How does Industrial IoT disrupt traditional manufacturing processes?

A) By resisting change and sticking with outdated methods.

B) By enhancing efficiency and driving innovation.

C) By hiding in the corner and avoiding work.

D) By creating chaos and confusion on the production line.

8) What's the Industrial IoT's favourite flavour in the oil industry?

A) Crude oil with a hint of data analytics.

B) Sweet success with a touch of process optimization.

C) Efficiency oil with a dash of innovation.

D) Chaos oil with a side of disruption.

9) How does Industrial IoT navigate the complex world of healthcare?

A) By avoiding hospitals and sticking with manufacturing.

B) By revolutionizing processes and enhancing efficiency.

C) By playing doctor and diagnosing illnesses.

D) By causing chaos and confusion in medical facilities.

10) What's Industrial IoT's ultimate goal in various industrial sectors?

A) To disrupt traditional practices and create chaos.

B) To enhance efficiency, optimize processes, and drive innovation.

C) To hide in the shadows and avoid attention.

D) To sabotage operations and increase downtime.

Summary

This Chapter delves into the realm of Industrial IoT (IIoT) applications and their profound impact on various industries. It emphasizes the fusion of advanced technologies with industrial processes, revolutionizing traditional setups with interconnected devices, sensors, and data analytics. The chapter begins by outlining the key components of IIoT, highlighting the critical role of sensors and devices in collecting real-time data to enable informed decision-making.

Furthermore, it discusses the significance of IIoT in transforming manufacturing processes, emphasizing operational productivity enhancements through real-time visibility and process optimization. The application of IIoT in healthcare is explored, showcasing its potential to improve patient care, reduce costs, and increase efficiency through remote monitoring and innovative healthcare solutions.

Inventory management, quality control, building automation, and safety are among the other domains where IIoT demonstrates its transformative capabilities. The chapter concludes with a recap of key IIoT application domains, including oil, chemical, and pharmaceutical industries, and discusses the profound impact of IIoT on the Fourth Industrial Revolution, marking a paradigm shift in business operations.

Abbreviation

Abbreviations	Terms
IIoT	Industrial Internet of Things
IoT	Internet of Things
HR	Human Resources
PPE	Personal Protective Equipment
AR	Augmented Reality
VR	Virtual Reality
R&D	Research and Development
CDO	Chief Digital Officer
CFO	Chief Financial Officer
CMO	Chief Marketing Officer

Glossary

1. Significance of Industry 4.0

- Programmable Logic Controllers (PLCs): Industrial digital computers for controlling manufacturing processes.
- Computer-Aided Design (CAD): Software for designing products.
- Computer-Aided Manufacturing (CAM): Software for manufacturing processes automation.
- Enterprise Resource Planning (ERP): Integrated management of core business processes.
- Industry 4.0: Automation of traditional manufacturing through innovative tech.

2. Introduction to IIoT

- Smart Devices: Devices with sensors and processors.
- Sensors: Devices detecting and responding to environmental inputs.
- IIoT: Industrial Internet of Things, connecting industrial devices.

2.2 Evolution of IIoT Technology

- Machine-to-Machine (M2M) Communication: Direct communication between devices.

2.3 Facilitating Data-Driven Decision-Making in Industries

- Supply Chain: Sequence of processes in production and distribution.
- Data Privacy: Protection of sensitive information from unauthorized access.

2.4 Influence on Job Roles and Skill Requirements

- Predictive Maintenance: Maintenance strategies predicting equipment failures.
- Actuators: Components controlling mechanisms or systems.

2.7 Creating New Business Models and Revenue Streams

- Product-Centric to Service-Centric: Shift from selling products to services.
- Predictive Maintenance, Remote Monitoring, and Asset Tracking: Strategies for maintenance and monitoring.

2.7.4 Leveraging Big Data Analytics for Actionable Insights

- Big Data Analytics: Analysis of large and complex datasets.

2.8 Addressing Cybersecurity Concerns

- Hacking, Data Breaches, and Malware Attacks: Unauthorized access and security breaches.

3. IIoT Network Technologies

- Haystack, SigFox: IoT network technologies.

3.4 IoT Networking Consideration and Challenges

- RESTful: Representational State Transfer, an architectural style.
- Universal Plug and Play (UPnP): Networking protocols for device discovery.
- Fog Computing: Distributed computing paradigm bringing computational resources closer to data.

3.6 Internet of Things Ecosystem

- Internet Protocol (IP): Rules for data transmission over networks.
- Licence-Free Areas: Geographic regions for wireless tech operation without licenses.
- IP Stacking or Non-IP Stacking: Techniques for organizing data packets.
- Packet Delivery Rate (PDR): Rate of successful data packet delivery.
- Ethernet, SCADA, Modbus TCP, DNP3, TCP/UDP: Networking and communication protocols.

5. Blockchain Technology in Industry 4.0

- Decentralized Autonomous Organizations (DAOs): Organizations with rules encoded as computer programs.

- Cryptographic Verification: Verification of digital messages or documents.

- Firewalls, Antivirus Software, Encryption: Security measures for protecting systems.

- Ransomware: Malicious software encrypting files for ransom.

- Decentralized Finance (DeFi): Financial services built on blockchain.

7. Cloud Empowering Industry 4.0, Revolutionizing Manufacturing and Beyond

- Array of Computing, Computing Frameworks: Diverse computing technologies and frameworks.

9. Significance of IIoT in the Industrial Sector

- Industrial IoT (IIoT): IoT in industrial settings.

9.8 Big Data

- Big Data: Large and complex datasets.

9.8.3.1 Data Security and Privacy

- GDPR: General Data Protection Regulation, data security and privacy regulation.

9.10.5.3 Computing Resources

- Computing Frameworks, Accelerative Computing: Computing resources and frameworks.

11. How to Progress IIoT Security in Industrial Settings

- AI and ML: Artificial Intelligence and Machine Learning for cybersecurity.

12.8.4 Product Development

- New Product Development (NPD): Process of bringing new products to market.

13.2.2 Limitations of Industry 3.0

- Cannibalism in Oligopolistic Markets: New products reducing demand for existing ones.

Reference

Industry 4.0

Definition: Refers to the ongoing automation of traditional manufacturing and industrial practices through innovative technologies. The convergence of the physical, digital, and biological spheres characterizes Industry 4.0.

Reference: https://www.linkedin.com/pulse/what-industry-40-very-easy-explanation-anyone-patrick-mutabazi-nrwze/

LEAN

Definition: LEAN places great emphasis on respecting and involving employees in decision-making processes, recognizing their expertise and contribution.

Reference: https://www.coursesidekick.com/management/4261816

Edge Computing

Definition: At the vanguard of the Industry 4.0 revolution stands edge computing, poised to redefine the very essence of data processing and analytics by introducing a distributed computing paradigm that brings computational resources closer to the point of data generation.

Reference: https://www.linkedin.com/pulse/supercomputer-market-expected-reach-lf7af

Smart Factory

Definition: The core of Industry 4.0 is the smart factory, where production processes, machinery, and products carry intelligence, providing information about themselves such as origin, production time, and location.

Reference: https://www.lpstech.com/site/en/insight/the-smart-factory-fusing-digital-transformation-in-manufacturing

Fourth Industrial Revolution

Definition: The ongoing automation and digitalization of traditional manufacturing practices using modern innovative technologies. Large-scale machine-to-machine communication and the Internet of Things are integrated for increased automation, improved communication, self-monitoring, and the production of intelligent products. This transformative revolution impacts all aspects of life from how we work to how we live.

Reference: https://www.thedrum.com/open-mic/cheat-sheet-what-is-industry-4-0-and-its-impact-and-opportunities-for-marketing

Internet of Things (IoT)

Definition: An interaction between the physical and digital worlds where the digital world interacts with the physical world using a plethora of sensors and actuators. IoT is defined as a paradigm in which computing and networking capabilities are embedded in any kind of conceivable object.

References:

- https://www.iaras.org/iaras/filedownloads/ijitws/2021/022-0004(2021).pdf
- https://www.hindawi.com/journals/jece/2017/9324035/
- https://www.linkedin.com/pulse/internet-things-pavithra-h

TCP/IP (Transmission Control Protocol/Internet Protocol)

Definition: TCP/IP underpins the internet, providing a simplified concrete implementation of layers in the OSI model, including application, presentation, session, transport, network, data link, and physical layers.

References:

- https://www.devx.com/terms/open-systems-interconnection-model/
- https://developer.ibm.com/articles/iot-lp101-connectivity-network-protocols/

LPWAN (Low-Power Wide-Area Network) Technologies

Definition: LPWAN technologies include LoRa (Long Range Physical Layer Protocol), Haystack, SigFox, LTE-M, and NB-IoT (Narrowband Internet of Things). According to Gartner, NB-IoT has become the standard for LPWAN networks.

Reference:

https://developer.ibm.com/articles/iot-lp101-connectivity-network-protocols/

LPWAN (Low-Power Wide-Area Network) Technologies

Definition: LPWAN technologies include LoRa (Long Range Physical Layer Protocol), Haystack, SigFox, LTE-M, and NB-IoT (Narrowband Internet of Things). It is used in low bandwidth situations, especially sensors and mobile devices on unreliable networks.

References:

- https://developer.ibm.com/articles/iot-lp101-connectivity-network-protocols/
- https://www.linkedin.com/pulse/12-common-iot-protocols-standards-ct-rf-antennas-inc

BLE (Bluetooth Low Energy)

Definition: BLE is a communication technology that allows devices to communicate with mobile apps.

Reference: https://www.locksmithledger.com/electronics-access-control/article/21116656/adding-an-electronic-layer

Internet History and Functionality

Information: Upload speeds may be slower than download speeds.

Reference: https://www.techbusinessnews.com.au/what-is-the-internet-how-does-it-work-and-whats-the-history/

Smart Factory

Definition: The core of Industry 4.0, where production processes and machinery are intelligent, and products carry information about themselves like what they are, when they were made, and where they are currently.

Reference: https://www.lpstech.com/site/en/insight/the-smart-factory-fusing-digital-transformation-in-manufacturing

Fourth Industrial Revolution

Definition: The ongoing automation and digitalization of traditional manufacturing practices using modern innovative technologies. Large-scale machine-to-machine communication and the Internet of Things are integrated for increased automation, improved communication, self-monitoring, and the production of intelligent products.

Reference: https://www.thedrum.com/open-mic/cheat-sheet-what-is-industry-4-0-and-its-impact-and-opportunities-for-marketing

Big Data

Diversity: Big data includes many types of data, including structured, semi-structured, and unstructured data. This difference creates problems in storage and identification.

Reference: https://www.linkedin.com/pulse/what-big-data-uses-types-analytics-interview-questions-jha

Internet of Things (IoT)

Definition: The Internet of Things is simply an interaction between the physical and digital worlds. The digital world interacts with the physical world using a plethora of sensors and actuators.

References:

- https://www.iaras.org/iaras/filedownloads/ijitws/2021/022-0004(2021).pdf/

- https://www.hindawi.com/journals/jece/2017/9324035//

- https://mrcet.com/downloads/digital_notes/EEE/IoT%20&%20Applications%20Digital%20Notes.pdf/

TCP/IP (Transmission Control Protocol/Internet Protocol)

Definition: TCP/IP underpins the internet and provides a simplified concrete implementation of these layers in the OSI model.

References:

- https://www.devx.com/terms/open-systems-interconnection-model/

- https://developer.ibm.com/articles/iot-lp101-connectivity-network-protocols/

LPWAN (Low-Power Wide-Area Network) Technologies

Definition: LPWAN technologies include LoRa, Haystack, SigFox, LTE-M, and NB-IoT. They are used in low bandwidth situations, especially sensors and mobile devices on unreliable networks.

References:

- https://developer.ibm.com/articles/iot-lp101-connectivity-network-protocols/
- https://www.linkedin.com/pulse/12-common-iot-protocols-standards-ct-rf-antennas-inc

"Digital Revolution: Industry 4.0," authored by the founder of Technoviz Automation Solutions Pvt Ltd, this comprehensive guide extends beyond theoretical insights, offering practical strategies for integrating technology and succeeding in the digital age.

For those seeking to harness the power of Industrial Automation, IIoT And Centre of Excellence (CoE), Technoviz Automation provides a range of services tailored to meet diverse needs

For Further details, you may contact us *www.technovizautomation.com,* Info@technovizautomation.com or +91 99997 65380.

Thank you…!